Daily Warm-Ups
PRE-ALGEBRA
NCTM Standards

Betsy Berry, Ph.D.

Indiana University, Purdue University, Fort Wayne

1 2 3 4 5 6 7 8 9 10
ISBN 978-0-8251-6460-6
Copyright © 2008
J. Weston Walch, Publisher
40 Walch Drive • Portland, ME 04103
www.walch.com
Printed in the United States of America

Table of Contents

Introduction

I taught mathematics for many years at the middle- and high-school level. I know how precious every moment can be in a class period. The problems, exercises, activities, reflections, and writing prompts in this collection of warm-ups are just what I wish I had during those years. These warm-ups are organized in seven parts around selected concepts designated by the National Council of Teachers of Mathematics *Principles and Standards for School Mathematics*: Number Sense and Number Theory; Measurement; Understanding Patterns, Relations, and Functions; Understanding Algebraic Symbols; Developing Algebraic Thinking Through and for Mathematical Modeling; Analyzing Change in Various Contexts; and Data Analysis and Probability. The NCTM process standards, especially communication and representation, are also integrated within these activities.

The warm-ups address all the concepts and skills that can be considered pre-algebra within NCTM. The NCTM expectations within each topic area are for grades 6 through 8. Some problems address more than one standard or expectation, but have been placed in the respective section according to the topic area that I thought was best addressed by the activity.

The warm-ups are organized by standards rather than by level of difficulty. Use your judgement to select appropriate problems for your students. The problems are not meant to be done in consecutive order from the beginning of the book to the end. Some of these problems are stand-alone, some can launch a topic, some can be used for journal prompts, and some refresh students' skills and concepts. All are meant to enhance and complement pre-algebra instruction. They do so by providing resources for teachers for those short spaces of time of 5 to 15 minutes when class time might go unused.

—Betsy Berry, Ph.D.

About the CD-ROM

Daily Warm-Ups: Pre-Algebra, NCTM Standards is provided in two convenient formats: an easy-to-use, reproducible book and a ready-to-print PDF on a companion CD-ROM. You can photocopy or print activities as needed, or project them on a large screen via your computer.

The depth and breadth of the collection gives you the opportunity to pick and choose specific skills and concepts that correspond to your curriculum and instruction. The activities address all of the skills and concepts in the NCTM Standards considered pre-algebra. Use the table of contents and the information in the section introductions to help you select appropriate tasks.

Suggestions for use:
- Choose an activity to project or print out and assign.
- Select a series of activities. Print the selection to create practice packets for learners who need help with specific skills or concepts.

Part 1: Number Sense and Number Theory

National Council of Teachers of Mathematics: "Instructional programs from pre-kindergarten through grade 12 should enable all students to understand numbers, ways of representing numbers, relationships among numbers . . . understand meaning of operations and how they relate to one another, and compute fluently and make reasonable estimates."

Expectations

- Work flexibly with fractions, decimals, and percents to solve problems.

- Relate and compare different forms of representation for a relationship.

- Understand and use ratios and proportions to represent quantitative relationships.

- Compare and order fractions, decimals, and percents efficiently and find their approximate locations on a number line.

- Use factors, multiples, prime factorization, and relatively prime numbers to solve problems.

- Develop meaning for integers, and represent and compare quantities with them.

- Develop, analyze, and explain methods for solving problems involving proportions, such as scaling and finding equivalent ratios.

Comparing Fractions

Jordan, Jose, and Brigit are comparing the sizes of the fractions $\frac{7}{8}$ and $\frac{2}{3}$. Jordan is using fraction strips, Jose is using graph paper, and Brigit is using decimal equivalences. Explain with pictures, diagrams, calculations, and words how each of them can demonstrate which value is larger.

Where Do They Go?

Using the number line below, show approximately where each number would fall. Explain your thinking.

1. $\sqrt{96}$

2. $\sqrt{35}$

3. $\sqrt{24}$

4. $\sqrt{17}$

Tearing and Stacking Paper

Mr. Andres poses the following problem to his math class: Take a large sheet of paper and tear it exactly in half. Then you have 2 sheets of paper. Put those 2 sheets together and tear them exactly in half. Then you have 4 sheets of paper. Continue this process of tearing and putting together for a total of 50 tears. If the paper is only $\frac{1}{1,000}$ of an inch thick, how many sheets of paper would there be? How thick or tall would the stack of paper be?

Bigger or Smaller?

Shanita and Leon are discussing their math homework. Leon says, "When you multiply one number by another number, the result is always bigger than the number you started with. When you divide one number by another number, the result is always smaller than the number you started with." Is Leon's statement always true? Can you give examples in which the statement is true and other examples in which the statement is false?

4

Ribbon and Bows

Veronica is making decorative bows for a craft project. She has 7 yards of velvet ribbon. Each bow requires $\frac{3}{4}$ of a yard of ribbon. How many bows will she be able to make with the ribbon she has? Will she have any ribbon left? Explain your thinking.

5

Dividing by Fractions

Write a word problem that could result in the number sentence below. Then explain how you know it fits the sentence.

$$5 \div \frac{2}{3} = 7R\frac{1}{2}$$

6

More Than One

Max is looking over his younger sister's math homework. He notices that she had written the following:

$$\frac{3}{4} + \frac{2}{3} = \frac{5}{7}$$

Without doing any calculations, he tells her that this is an incorrect math sentence. Is he right? What might he be thinking? How could he explain his reasoning to his sister?

Birthday Roses

Alonzo is planning to purchase roses for his mother on her birthday. He has seen them advertised at 12 roses for $15.00 and 20 roses for $23.00. Which is the better buy?

8

Balancing a Milk Bottle

An American named Ashrita Furman holds more Guinness World Records than any other person. In April 1998, he walked 81 miles in 23 hours, 35 minutes while balancing a milk bottle on his head. How fast did he walk in miles per hour?

Stacking T-shirts

A clerk in a sportswear department was asked to arrange T-shirts in a display in stacks of equal size. When she separated the T-shirts into stacks of 4, there was 1 left over. When she tried stacks of 5, there was still 1 left over. The same was true for stacks of 6. However, she was able to arrange the shirts evenly in stacks of 7. How many T-shirts were in the display that she was arranging?

10

Perfect Numbers

A number is said to be perfect if it is equal to the sum of all its proper factors. Proper factors are all the factors of the number including 1, except for the number itself. Find three numbers that are perfect and show why they are perfect numbers using the definition given.

What Day of the Week?

Saari and Anatole are cooped up inside because of bad weather. Their mother, a middle-school teacher, poses a puzzler for them. She says that Anatole was born 759 days after Saari. If Saari was born on a Wednesday, on what day of the week was Anatole born? Find the day of the week and explain your strategy for finding it.

12

Greatest Common Factor

Find the greatest common factor (GCF) of 360, 336, and 1,260. Describe the strategy you used to find the GCF. Will your strategy work for all situations?

13

Library Day

Marika, Bobbie, and Claude all visit their local public library on a regular schedule. Marika visits every 15 days, Bobbie goes every 12 days, and Claude goes every 25 days. If they are all at the library today, in how many days from now will they all be there again? What strategy did you use to find the number of days? What does your answer represent?

14

Cicada Cycles

Stephan's grandfather told him that some cicadas have 13-year or 17-year cycles. He said that one year, the cicadas were so numerous on his family farm that they ate all the crops. Stephan guessed that perhaps both the 13-year and the 17-year cicadas came out that year.

1. If we assume Stephan is correct, how many years will have passed when the 13-year and 17-year cicadas come out together again?

2. Imagine that there are 12-year, 14-year, and 16-year cicadas, and they all come out this year. How many years will pass before they all come out together again?

3. Explain how you got your answer. What number does your answer represent?

15

Make a True Sentence

For the sentence below, insert mathematical symbols of any kind to make the sentence true. It is possible to make 2- or 3-digit numbers by not inserting any symbols between the numbers (for instance, by putting 5 and 6 together to make 56).

$$1 \quad 2 \quad 3 \quad 4 \quad 5 \quad 6 \quad 7 \quad 8 \quad 9 = 100$$

16

A Number Puzzle

Choose any number. Add the number that is 1 more than your original number. Add 11. Divide by 2. Subtract your original number. What is your answer? Do the puzzle again for other numbers. Why do you get the answers that you get? Will this always work?

Rectangular Configurations

Washington Center Junior Chamber of Commerce is hosting a holiday craft show. They are renting spaces in a variety of rectangular configurations with side lengths that are whole numbers. The spaces are between 30 and 50 square yards.

1. Which number(s) of square yards, between 30 and 50, give the most choices for rectangular arrangements? Why?

2. Which number(s) of square yards, between 30 and 50, allow for square spaces to rent? Why? Explain your thinking.

18

Numbering Pages

Dylan used 2,989 digits to number the pages of a book. How many pages are in the book? Remember that digits are the symbols 0, 1, 2, 3, 4, 5, 6, 7, 8, 9. Explain how you got your answer.

Planting Evergreen Trees

Garrett works on a tree farm where they just received a shipment of 144 seedlings. He wants to plant rows of trees in a rectangular pattern. How many different rectangular configurations can he make? What are they? Explain your thinking.

20

Volleyball Tournament

Every year in March, Maple Hill Middle School holds intramural activities for the whole school. This year, Sam and Sarah are organizing a volleyball tournament. Nine teams have registered to play. If each team plays every other team once, how many games must Sam and Sarah schedule?

21

Using 10 x 10 Grids 1

Adriana has shaded a portion of each 10 × 10 grid pictured below. What fraction of each has been shaded? Express the fraction as a decimal. What percent of each grid has been shaded?

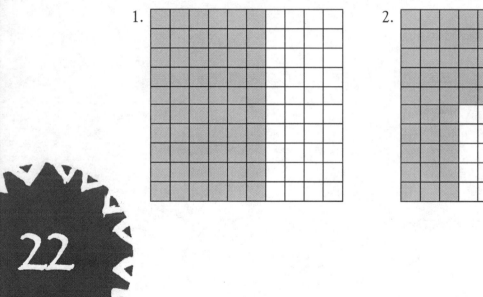

1.

2.

22

Using 10 x 10 Grids II

Micaela is learning about the connection between decimals and fractions. She has made the following diagram in her notes to represent multiplying decimals less than 1. What multiplication sentence(s) might be represented by the darker shaded area of her drawing?

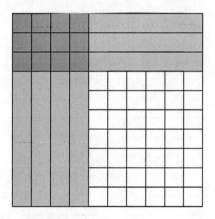

Square Pizza

Lin, Lon, Lu, and Lau have ordered a pizza from the Tip Top Pizza Palace. The pizzas only come in one size and are in the shape of a square. Lau has just had a blueberry smoothie and is not very hungry, but she thinks that she might eat 10% of the pizza. Lon is famished and thinks that he might eat half of the pizza. Lin thinks she might eat about 35% of the pizza, and Lu thinks he might eat 15%. If the four friends eat the portions that they have predicted, what percent of the pizza will remain? Justify your thinking using a 10 × 10 grid like the one below.

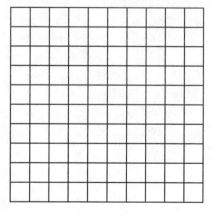

Cookie Sales

The seventh-grade class has been having a bake sale every Friday. One Friday, the students noticed that many oatmeal cookies priced at $0.75 each were not selling. They decided to lower the price and then sold all the cookies for a total of $31.93. How much did they charge for each cookie, and how many did they sell?

25

Baking Blueberry Pies

Mrs. Berry is famous for her pies. She has been making pies in her bakery for many years. She knows that it takes $1\frac{2}{3}$ cups of flour to make her special pie crust. She buys flour in 25-pound bags and knows that each pound contains about 3 cups of flour. How many pies can she expect to make from a 25-pound bag of flour?

26

Fraction Sense

1. Name two fractions that come between $\frac{3}{5}$ and $\frac{4}{5}$. Justify your answer in two or more different ways.

2. Which of the following fractions is larger: $\frac{8}{9}$ or $\frac{11}{12}$? Explain how you can determine which fraction is larger without changing the fractions to decimals or finding a common denominator.

27

Super Sub Sandwich

The students at Abigail Adams Middle School tried to make the biggest sub sandwich on record. They made a sandwich that was $12\frac{3}{4}$ feet long. After they were told that this did not break the record, they decided to divide the sandwich into smaller portions and share it with other students.

1. If each portion was $\frac{1}{2}$ foot long, how many students would get a portion?

2. If each portion was $\frac{3}{4}$ foot long, how many students would get a portion?

3. Explain and illustrate your thinking using a diagram.

28

Working with Integers I

Using appropriate arrow notations on number lines such as the one pictured below, represent the expressions given and find the results. Use a new number line for each sentence.

1. $-8 + 12 = ?$

2. $(-7) + (-3) = ?$

3. $5 + (-9) = ?$

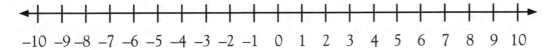

29

Working with Integers II

Look at each diagram below. Write a true sentence showing the integer calculations pictured.

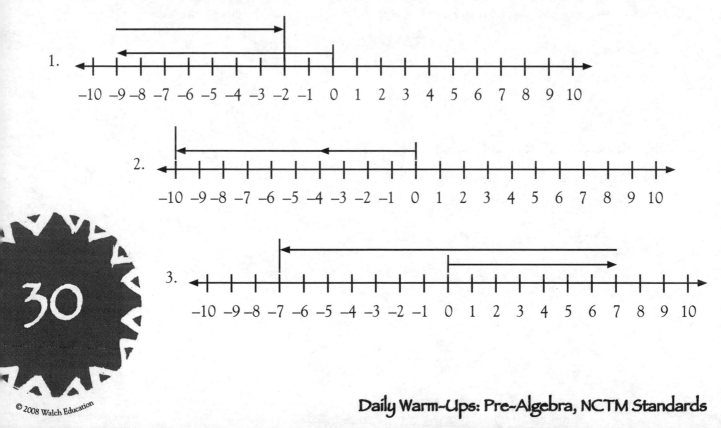

1.

2.

3.

30

Chip-Board Integers 1

Andrew and Brittany are exploring integers using several different models. They are drawing representations using black and white circular chips. The white chips represent positive numbers. The black chips represent negative numbers. Write a number sentence to symbolize each set of chip boards that they have drawn.

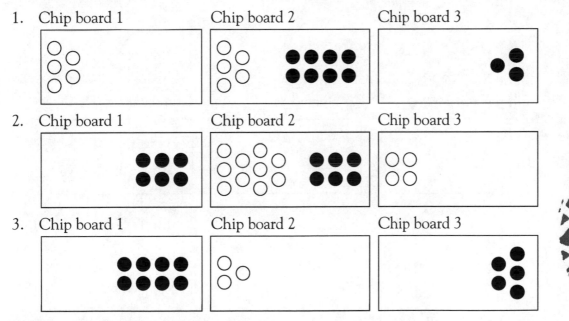

1. Chip board 1   Chip board 2   Chip board 3

2. Chip board 1   Chip board 2   Chip board 3

3. Chip board 1   Chip board 2   Chip board 3

31

Chip-Board Integers II

Diagram and solve the following integer operations. Use chip boards such as the ones pictured below. Use black circles, or chips, to represent negative values. Use white circles, or chips, to represent positive values. Use a series of two or three chip boards for each sentence.

1. $(-5) + (9) = ?$

2. $12 + (-7) = ?$

3. $(-8) - (-5) = ?$

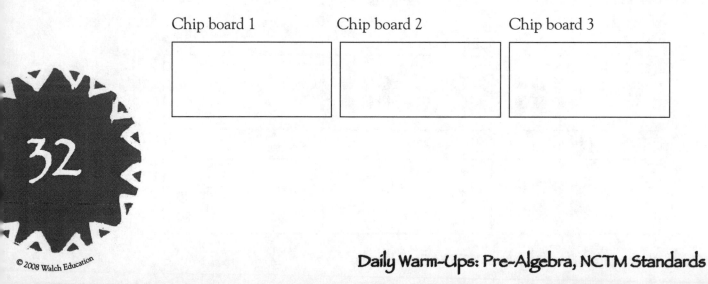

Chip board 1   Chip board 2   Chip board 3

Chip-Board Integers III

Explain how you might show the operation $(-6) - (-12) = ?$ on a chip board or a series of chip boards such as the ones pictured below. Use black circles, or chips, to represent negative values. Use white circles, or chips, to represent positive values. A black chip and a white chip together represent zero.

Chip board 1

Chip board 2

Chip board 3

Integer Practice

Find the missing value in each sentence below.

1. $(-6) + (-12) = ?$

2. $25 - (-8) = ?$

3. $? + 17 = (-4)$

4. $11 + ? = 6$

5. $\dfrac{3}{4} - ? = 1$

6. $(-7) + ? = 0$

7. $2\dfrac{1}{2} + ? = \dfrac{1}{2}$

8. $? - (-5) = 0$

9. $-3.5 - \dfrac{7}{2} = ?$

10. $(-5.6) - ? = (-3.4)$

11. $? + 7\dfrac{2}{5} = 3\dfrac{3}{5}$

12. $-\dfrac{87}{10} + ? = 10$

34

Daily Warm-Ups: Pre-Algebra, NCTM Standards

Weighing Coins

Denae and Lindsay are working on a puzzle. Their mother gave them 8 coins that look exactly alike. The only difference is that 1 coin is heavier than the other 7. Their mother also gave them a balance scale like the one shown below. She tells Denae and Lindsay to find the heaviest coin in as few weighings as possible. What are the fewest number of weighings they need to find the heaviest coin? Explain your answer.

A Circle of Students

Every year on the first day of spring, all the students at Thoreau Middle School stand in a perfect, evenly spaced circle on the athletic field in honor of the vernal equinox. This year, Taylor notices that she is in seventh place on the circle. Directly opposite her in position number 791 is her friend Mattie. How many students make up the circle this year?

36

Four 4s

Express each of the numbers 1 through 10, using only the addition, subtraction, multiplication, and division symbols; four 4s; and parentheses, if necessary.

37

Sharing Cider

Lola and Charlotte have purchased an 8-gallon container of cider at the local farmers' market. They would like to divide the cider evenly between them. However, they only have a 5-gallon container and a 3-gallon container. How can they use just these containers to solve their problem?

38

Chess Challenge

Sybil, Chuck, and Theodore have just finished playing three games of chess. There was only one loser in each game. Sybil lost the first game, Chuck lost the second game, and Theodore lost the third game. After each game, the loser was required to double the money of the other two friends. After the three games, each player had exactly $24. How much did each person start with?

Population Density

Evan, who is from Indiana, and Carlos, who is from Wyoming, met at basketball camp last summer. They made comparisons of their two states using population and land area. Find the answers to some of the questions that they posed using the information in the chart.

State	2000 population	Land area in square miles
Indiana	5,544,159	35,870
Wyoming	453,588	97,104

1. What is the population density for each state?

2. How many people would have to leave Indiana for it to have the same population density as Wyoming?

3. How many people would have to move from Indiana to Wyoming to give the two states the same population density?

40

Part 2: Measurement

National Council of Teachers of Mathematics: "Instructional programs from pre-kindergarten through grade 12 should enable all students to understand measurable attributes of objects and the units, systems, and processes of measurement . . . and apply appropriate techniques, tools, and formulas to determine measurements."

Expectations

- Understand relationships among units and convert from one unit to another within the same system.

- Understand, select, and use units of appropriate size and type to measure angles, perimeter, area, surface area, and volume.

- Develop and use formulas to determine the circumference of circles and the area of triangles, parallelograms, trapezoids, and circles, and develop strategies to find the area of more complex shapes.

- Develop strategies to determine the surface area and volume of selected prisms, pyramids, and cylinders.

- Solve problems involving scale factors, using ratio and proportion.

- Solve simple problems involving rates and derived measurements for such attributes as velocity and density.

Buying Carpet

1. Todd and Maria are renovating their home. Their living room is a rectangle that measures 15 feet wide and 21 feet long. They are looking at new carpet that is on sale for $7.32 per square yard. How much will it cost to install this carpet in their living room?

2. They also want to install baseboard molding in the living room after the carpeting is done. The molding they want costs $134.00 per 8-foot piece. How much should they buy, and what will it cost?

41

Blob Area

Zoe has created the picture below using her computer. Her grid consists of squares with 1-inch sides. Estimate the area of the blob Zoe has drawn on her grid.

42

Pentominoes

Pentominoes are shapes made of 5 adjacent squares on a grid or 5 tiles connected along their edges. For example, the shaded area in the grid below represents a pentomino.

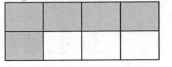

However, the figure below does NOT represent a pentomino.

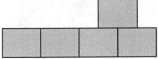

A pentomino that has been flipped or turned does not result in another pentomino. For example, the pentomino below is thought of as the same pentomino as the one shown above.

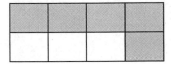

Use grid paper or tiles to find all the pentomino shapes that you can. How many are there? What are the perimeters of the pentominoes that you find?

Sketching Rectangles

Using grid paper, sketch all the rectangles you can find that have an area of 48 square units. How many different rectangles did you find? What is the perimeter of each rectangle? Which rectangle has the smallest perimeter? Which has the largest? Did you find a rectangle that is a perfect square? Why or why not?

44

An Unusual Area

Vachon is helping his uncle tile a patio that has an unusual shape. He has drawn a sketch on grid paper to represent the area that they need to cover. Each unit on the grid paper represents 2 feet. Find the area of the figure, and explain your strategy.

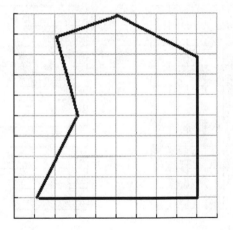

Vanishing Wetlands

Bonita works for the Desoto Park Service. The Little Otter Wetland Area that she monitors has had drought conditions recently. She is preparing a report on the drought for the park service. The grid below represents a model for the change in area of the wetland. What was the percent change in wetland area from August 2005 to August 2007? Explain your thinking.

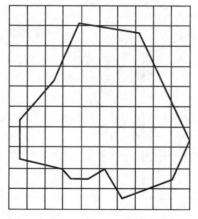

August 2005 model

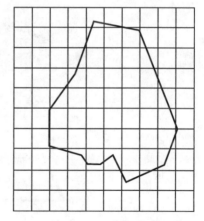

August 2007 model

46

Considering Pizza Costs

Paisano's Pizza Parlor has just opened up across the street from Mei's apartment. Mei is thinking about inviting some friends over for pizza and ordering from Paisano's. She wants to get the most pizza for her money. The menu shows the following prices:

6-inch round pizza...................$7.50
12-inch round pizza.................$12.00
18-inch round pizza.................$16.00

1. How many 6-inch pizzas have the same amount of pizza as one 12-inch pizza?

2. How many 6-inch pizzas have the same amount as one 18-inch pizza?

47

Pondering Pizza Prices

Antonio is the new owner of Paisano's Pizza Parlor. He wants to change the way his pizzas are priced. He is thinking about pricing them according to the diameter of each pizza. His cousin Vinnie suggests that he price them according to the size of the circumference of each pizza. His wife Bianca recommends pricing by the area of each pizza. Which method would you recommend to Antonio? Explain your reasoning.

48

Area and Perimeter

The figure below represents a rectangle with a length of 48 centimeters and a height of 16 centimeters. A semicircle with a radius of 8 centimeters has been cut out of each end of the rectangle. Find the area and perimeter of the shaded region of the figure.

49

Yankee Stadium Tree Replacement

In New York City, a new stadium for the Yankees baseball team is being built. During the planning process for the new ballpark, a concern was raised about how many trees would be lost during construction of the new site. The Parks Department estimated that 165 trees could be protected, but that 377 trees would be lost. The New York Supreme Court ordered that the removed trees be replaced using the Parks Department "basal area tree replacement formula." This means that a total of approximately 592 square feet of area must be replaced by new trees. If the diameter of each new tree averages 3.5 inches, how many new trees will the Parks Department need to plant?

50

Areas of Triangles

Give the base, the height, and the area of each triangle pictured below. Are the triangles congruent? Explain your thinking.

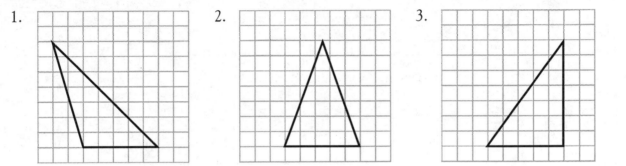

1.

2.

3.

51

Paolo's Pizza Pricing

Paolo has just started working at his Uncle Antonio's pizza parlor. He is trying to figure out which size meat lover's pizza provides the best value.

Meat Lover's Special

9-inch round pizza……………….$10.50	
12-inch round pizza……………..$15.00	
18-inch round pizza……………..$19.00	

Which pizza listed on the menu provides the best value? Write a few sentences that explain your reasoning.

52

Circles and Squares

Find the area of the shaded region for each figure below.

1.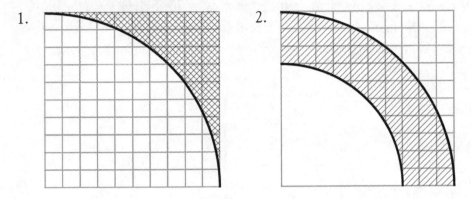

2.

53

Yearbook Ads

Bailey and Inez are in charge of advertising for the Benjamin Franklin Middle School yearbook. A full-page ad in the yearbook sells for $260 and covers an area of 8 inches by 10 inches. One page of ads has a layout like the one shown in the figure below. What fraction of the total page is each section shown in the figure? How much should an advertiser be charged for each space?

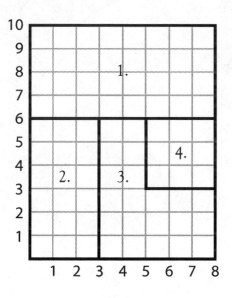

Storing DVDs

Malik wants to make an open-top cardboard box to fit on a certain shelf in his closet to store his DVD collection. He has drawn the sketch below of the box that he needs.

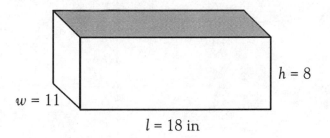

$w = 11$

$h = 8$

$l = 18$ in

1. How much cardboard will Malik need to create this box?

2. If a typical DVD package measures 7.5 inches × 5.25 inches × 0.5 inches, how many DVDs can Malik expect to store in the box he makes?

3. Malik's dad has brought him a piece of cardboard that measures 2 feet by 3 feet. Is this enough cardboard for the project? Could Malik cut all the pieces that he needs? Explain why or why not.

Daily Warm-Ups: Pre-Algebra, NCTM Standards

Comparing Cylinders

Two cylinders are pictured below. All the dimensions of cylinder A are 3 times the dimensions of cylinder B.

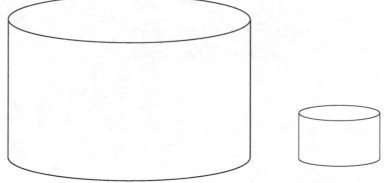

1. What is the ratio of the radius of cylinder A to the radius of cylinder B?

2. What is the ratio of the height of cylinder A to the height of cylinder B?

3. What is the ratio of the surface area of cylinder A to the surface area of cylinder B?

4. What is the ratio of the volume of cylinder A to the volume of cylinder B?

56

Daily Warm-Ups: Pre-Algebra, NCTM Standards

Compressing Trash

According to the U.S. Environmental Protection Agency's 2003 report, every person generates an average of 4.6 pounds of garbage per day. Of this, approximately 25% is recycled. If a cubic foot of compressed garbage weighs about 50 pounds, how much space would be needed for the garbage generated by a family of four in one year?

Storing Trash

According to the U.S. Environmental Protection Agency, U.S. residents, businesses, and institutions made more than 236 million tons of solid municipal waste in 2003. Every ton of recycled material saves 2.5 cubic yards of landfill space. Andrew and Todd are wondering how many math classrooms like theirs would be needed to store all that garbage. Their classroom is 45 feet long, 30 feet wide, and 12 feet high. How many rooms would be needed to store the garbage?

58

Concession Concern

Emile and Sharonda are in charge of concessions for their school's football games. They usually sell popcorn in cylindrical containers like the one below on the right. However, they have just noticed that their supply of these containers is gone, and there's no time to get more before the next game. Emile notices that there are plenty of the cone-shaped containers like the one on the left. He suggests that they use those. Both of the containers have a height of 8 inches and a radius of 3 inches.

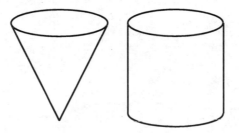

1. What is the difference between what the cone will hold and what the cylinder will hold?

2. If Emile and Sharonda usually charge $1.50 for the cylinder full of popcorn, what should they charge for the cone?

59

Popcorn Pricing

Twin sisters Elizabeth and Emma enjoy going to the movies on Saturday afternoons. Sometimes they each buy a small popcorn for $3.00, and sometimes they buy a large popcorn that costs $6.00 and share it. Both containers are cylinders. The heights of the two containers are the same. However, the radius of the large container is about twice the radius of the small container. Which purchase gives the sisters the most popcorn for their money? Draw a sketch of the two popcorn containers, and explain your thinking.

60

Juice Packaging

The Johnny Appleseed Juice Company is changing its packaging. The juice currently comes in a cylindrical container like the one pictured below on the right. The company is considering changing this container to a prism. The prism will have a square base that has the same width as the diameter of the cylinder, which is 5 inches. The height of both containers is $7\frac{1}{2}$ inches.

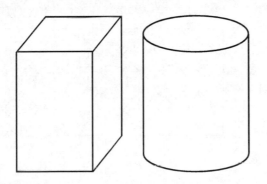

1. What is the volume of each container?

2. If the company was charging $3.29 for the cylindrical container of juice, what would be a fair price to charge for the new container?

Collette's Chocolates

Collette works as the marketing manager for a small chocolate-making company. She has data that shows that the company's orange creams and French mints are the most recent best sellers. The French mints are a little more popular than the orange creams. Her supervisor has asked her to create new packaging for a Valentine's Day sales promotion. Collette has come up with the following plan for different-sized boxes of chocolates. The round shapes represent orange creams. The squares represent French mints.

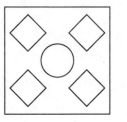

2 × 2 box

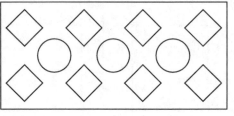

2 × 4 box

1. If Collette continues her pattern and she increases both the number of rows and/or the number of columns, what would a 4 × 6 box of chocolates look like? Draw several more boxes of chocolates that you think will continue Collette's design.

2. How many orange creams would be in the 4 × 6 box? Write a rule for finding the number of orange creams if you know the rows and columns of French mints.

Daily Warm-Ups: Pre-Algebra, NCTM Standards

Raising Zebra Finches

Kanya and her father raise and breed zebra finches. They have an aviary made of wood and wire mesh that measures 6 feet wide by 4 feet deep by 5 feet high. Their manual says that every pair of finches should have 3 to 4 square feet of area. The height of the space may vary. Kanya and her father now have 5 pairs of finches in the cage. They would like to expand their flock to a total of 12 pairs of birds. Do they have enough space for the 5 pairs they have now? They want to keep their finches happy and have decided to provide the maximum space for them. How could they increase the size of the cage to make space for 12 pairs? How much more wire mesh will they need to enclose their new aviary? (They do not put the wire mesh on the floor of the cage, but they cover all other sides with it.)

63

Perimeter Expressions

Rajah is working on her math homework. She has been given the sketch below, where r represents both the radius of the semicircle and the height of the rectangle that the semicircle is attached to. She is trying to decide on an expression that will represent the perimeter of the figure. She has written out the expressions below. Do they correctly represent the perimeter of the shape? Explain why or why not for each.

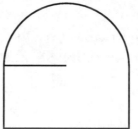

1. $r(2 + \pi)$

2. $4r + \pi r$

3. $r + 2r + r + 2\pi r$

4. $r(4 + \pi)$

5. $2r + \pi r$

64

Cylinder Expressions

The cylinder below has a base area of 45 cm². It is partially filled with liquid up to a height of x cm. The height from the top of the liquid to the top of the cylinder is represented by y cm. The total height is represented by h cm. Which expressions below properly represent the total volume of the cylinder?

a. $45(h - x)$

b. $45(x + y)$

c. $45x + 45y$

d. $45xy$

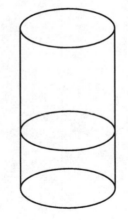

Cynthia's Cylinder

Cynthia has drawn the model below to represent a cylindrical cardboard box that she needs to package a gift of lotion for her mother. If the package needs to be 8 inches tall and have a diameter of 3 inches, what is the area of the rectangle pictured? What is the total surface area of the box that she will create from her design?

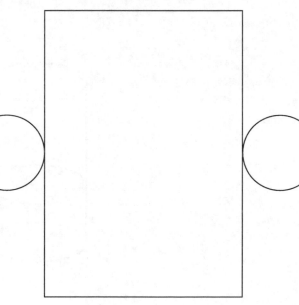

66

Beach Volleyball

Beach volleyball is played on beaches around the world and is a growing sport in the United States and Canada. The ball used for beach volleyball is a little larger in diameter than a traditional volleyball. It is usually brightly colored compared to the white ball used in indoor games. The Victor Volleyball Company packages its individual beach volleyballs in display cartons that measure 1 foot on each edge. It then ships 12 boxed volleyballs per carton to sporting goods stores all over the world.

1. Find the dimensions of all the different possible shipping cartons that the Victor Volleyball Company could use for exactly 12 balls.

2. Find the surface area of each shipping carton.

3. Which carton needs the least amount of cardboard?

4. Imagine that the Victor Volleyball Company has a greater demand for volleyballs and decides to package 24 boxed volleyballs in a carton. How much packaging material will the company need to create the box with the least material? How much more material is this than the least amount needed for shipping 12 balls?

67

Volume and Surface Area 1

The drawing below is a flat pattern or a net. When folded, it creates a box in the shape of a rectangular prism. Draw a sketch of the box and determine its total surface area and volume.

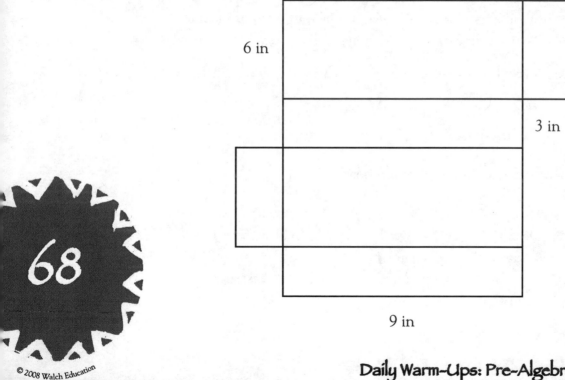

6 in

3 in

9 in

68

Volume and Surface Area 11

The drawing below represents an open-topped rectangular prism with a length of 13 feet, a height of 6 feet, and a width of 5 feet. Draw a flat pattern or a net for the figure. Determine the surface area and volume of the box.

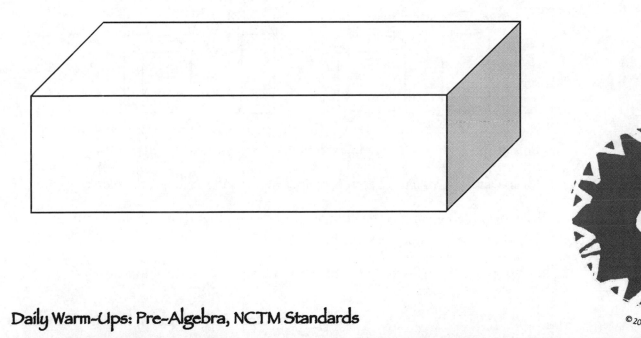

69

Paper Cylinders

Erica has a sheet of paper that is $8\frac{1}{2}$ inches by 11 inches. Without cutting the paper, she wants to make a container with the greatest possible volume. (She will make the top and bottom of the container with another sheet of paper.) She thought of rolling the paper to make an open-ended cylinder and realized that there are two ways to do this. Her friend Aleah suggests folding the paper to make a rectangular prism with square ends instead. Erica points out that there are also two ways to fold the paper to make the sides of a prism with a square base.

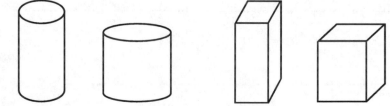

1. Predict which of the four containers has the greatest volume. You may want to make models of the containers to help explain your reasoning.

2. Now find the volume of each container. What is the volume of the largest container?

3. How much greater is the volume of that container than the volume of the other container of the same height?

4. Write a note to Erica that explains and justifies your thinking clearly.

70

A Rectangular Box

1. Make a sketch of a rectangular box with a base of 3 inches by 5 inches and a height of 7 inches. How many unit cubes would fit in a single layer at the bottom of the box?

2. How many identical layers of unit cubes could be stacked in the box?

3. What is the volume and surface area of the box?

Tennis Ball Packaging

Rochelle works as a packaging engineer for the Creative Carton Company. The company wants to remove the air from containers of tennis balls so the balls will retain a good bounce. Rochelle needs to know how much air there is in a standard container of tennis balls. Help Rochelle by finding the amount of empty space in a cylindrical container that is 18 centimeters tall and contains 3 tennis balls that are 6 centimeters in diameter. Then write a note to her about how you found your answer.

72

Road Trip

1. Two cars leave the same parking lot in Centerville at noon. One travels due north, and the other travels directly east. Suppose the northbound car is traveling at 60 mph and the eastbound car is traveling at 50 mph. Make a table that shows the distance each car has traveled and the distance between the two cars after 1 hour, 2 hours, 3 hours, and so forth. Describe how the distances are changing.

2. Suppose the northbound car is traveling at 40 mph, and after 2 hours the two cars are 100 miles apart. How fast is the other car going?

3. Draw a diagram to help explain the situation. Explain your thinking clearly.

73

Tree Dilemma

Eric's neighbor wants to cut down a dead tree that is in his yard. Eric is worried that when the tree is cut, it will fall on his garage, which is 42 feet from the tree. His neighbor decides to measure the height of the tree by using its shadow. The tree's shadow measures 47.25 feet. At the same time, Eric puts a yardstick next to the tree, and the yardstick casts a shadow of 3.5 feet. Will the tree hit Eric's garage if it falls the wrong way? Explain carefully. Include a sketch of the situation to help clarify your thinking.

74

Part 3: Understanding Patterns, Relations, and Functions

National Council of Teachers of Mathematics: "Instructional programs from pre-kindergarten through grade 12 should enable all students to understand patterns, relations, and functions."

Expectations

- Represent, analyze, and generalize a variety of patterns with tables, graphs, words, and, when possible, symbolic rules.

- Relate and compare different forms of representation for a relationship.

- Identify functions as linear or nonlinear and contrast their properties from tables, graphs, or equations.

Border Tiles I

Landscapers often use square tiles as borders for garden plots and pools. The drawings below represent square pools for goldfish surrounded by 1-foot square tiles. For example, if the square pool is 2 × 2, there are 12 tiles in the border. If the square pool is 3 × 3, there are 16 square tiles in the border.

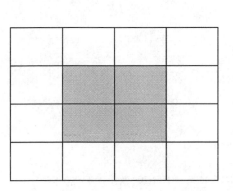
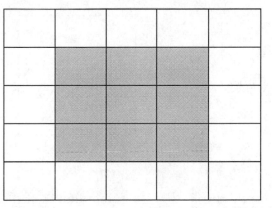

Collect data for a number of different-sized square pools. How many square tiles would be in the border around a pool that is 5 × 5? A pool that is 10 × 10? What patterns do you see in your data? Show and explain your thinking. Graph the data that you have collected. Then describe your graph.

Daily Warm-Ups: Pre-Algebra, NCTM Standards

75

Border Tiles II

Landscapers often use square tiles as borders for garden plots and pools. The drawing below represents a square pool for goldfish surrounded by 1-foot square tiles.

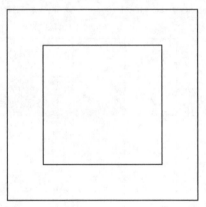

Maria thinks that the total number of tiles around the outside of the pool could be represented by $(n + 2) + (n + 2) + n + n$. Carlos thinks that he can represent the total number of tiles by $4(n + 1)$. What do you think about Carlos and Maria's representations? How do you think they would explain their thinking? Can you show if their expressions are equivalent or not?

Border Tiles III

Landscapers often use square tiles as borders for garden plots and pools. The drawing represents a square pool for goldfish surrounded by 1-foot square tiles.

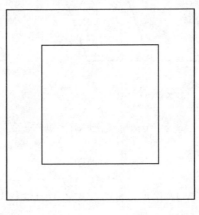

1. How many tiles will be needed for the border of this pool with an edge of length S feet?

2. Express the total tiles in as many different ways as you can. Be ready to explain why your different ways are equivalent.

Polly's Popsicle-Stick Parallelograms

Polly is creating patterns with Popsicle sticks.

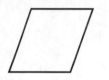

first figure

4 Popsicle sticks

second figure

7 Popsicle sticks

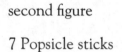

third figure

10 Popsicle sticks

1. How many Popsicle sticks are in the thirteenth figure in her pattern?

2. Write both an explicit equation and a recursive equation for the number of Popsicle sticks t_n for the nth figure in Polly's pattern.

3. What would a figure that has exactly 71 Popsicle sticks look like?

78

Daily Warm-Ups: Pre-Algebra, NCTM Standards

Triangular Numbers

Triangular numbers can be represented using dots, as shown below. The first triangular number is 1, the second is 3, the third is 6, and the fourth is 10, as pictured. How many dots will be in the fifth and sixth figures? Can you predict how many dots will be in the nth figure? Write an equation that could be used to determine the nth figure.

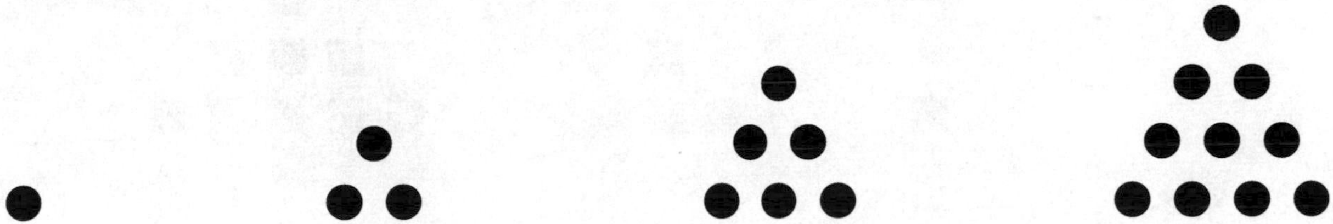

Patterns with Squares

Alexandra is babysitting her little brother. To entertain him, she is building structures with his blocks. The picture below represents three of the structures that she has built.

1. Sketch pictures of what you think the next two structures will look like if Alexandra continues the pattern that she has started. What number pattern is suggested by the number of blocks used in this sequence of structures?

2. How many blocks would be in the tenth structure if the pattern continues?

3. How many blocks would be in the 100th structure?

4. How many blocks would be in the *n*th structure?

80

Pentagonal Numbers

Pentagonal numbers can be represented using circles to form a pentagonal array as in the diagram below. Find a generalization for the nth pentagonal number.

 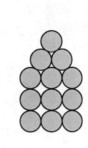 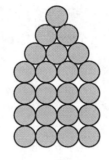

P_1 $\qquad$ P_2 $\qquad$ P_3 $\qquad$ P_4

81

Newspaper Numbers

Many newspapers consist of large sheets of paper that are folded in half to form the pages. For example, a newspaper that is made from 2 large sheets of paper would have 8 pages. Consider the following questions about newspapers created in this way.

1. If there were 10 sheets of paper, how many pages would there be?

2. What page numbers would appear in the center, or on the innermost sheet of the newspaper? What is the sum of the numbers?

3. What is the sum of the page numbers that would appear on ANY one side of ANY sheet of the newspaper?

4. What is the sum of all the pages of the newspaper?

5. Write a rule in words or symbols that would give the sum of the pages for a newspaper with N sheets of paper.

82

Stacking Cereal Boxes

Freya works part-time in her family's grocery store. Her dad has asked her to stack 45 cereal boxes in a display area for an upcoming sale. The boxes have to be stacked in a triangle with 1 fewer box in each row ending with 1 box at the top (like the sample stack shown below). How many boxes should Freya put in the bottom row? Suppose she had B boxes to stack in the display. Help her find a rule or formula to determine how many boxes she should put in the bottom row.

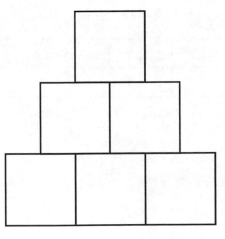

Pizza Cutting

Paolo loves pizza. He even dreams of pizza and makes up math problems about pizza. He wonders how many pieces of pizza he could get by cutting a very large circular pizza exactly 7 times. He decides that it doesn't matter if the pieces are the same size. He also decides that it isn't necessary for the cuts to go through the center of the pizza. How many pieces of pizza could he make? What if Paolo made n cuts in the pizza?

What is the greatest number of pieces he could make? Use the table below to help you.

Number of cuts	0	1	2	3	4	5	6	7	8	9			n
Number of pieces	1	2	4	7									

84

An Odd Pyramid

Look for patterns in the pyramid below. Then answer the following questions.

$$1$$
$$3 \qquad 5$$
$$7 \qquad 9 \qquad 11$$
$$13 \qquad 15 \qquad 17 \qquad 19$$
$$21 \qquad 23 \qquad 25 \qquad 27 \qquad 29$$

1. If the pyramid continues, what will be the middle number in row 25? In row 50? In row *n*?

2. What is the difference between the first and last number in row 4? In row 5? In row 40? In row *n*?

3. What is the last number in row 5? In row 6? In row *n*?

4. What is the sum of the numbers in row *n*?

5. What is the sum of all the numbers *through* row *n*?

Daily Warm-Ups: Pre-Algebra, NCTM Standards

85

How Long? How Far?

The Carmona family lives in Minnesota. They are driving to Florida for a vacation at an average speed of 60 miles per hour. Write an equation for a rule that can be used to calculate the distance they have traveled after any given number of hours. Then write a brief letter to the Carmona family that describes the advantages of having an equation, a table, and a graph to represent their situation.

86

Game Time

Abby and her brother Harry like to play a game called "U-Say, I-Say." Harry gives the "U-Say" number (an integer between −10 and +10). Abby has a secret rule she performs on the number that results in the "I-Say" number. Complete the table below by giving the missing "I-Say" values. Then describe Abby's rule in words and symbols.

U-Say	3	0	−4	1	2	5
I-Say	11	2	−10	5		

87

Table Patterns I

Look at the tables below. For each table, do the following:

a. Describe symbolically or in words the pattern hiding in the table.

b. Give the missing values in the table.

c. Tell whether the relationship between x and y in the table represents a quadratic, a linear, an inverse, an exponential, or other relationship. Explain how you know.

1.

x	0	1	2	3	4	5
y	1	3	9	27		

2.

x	−1	0	1	2	3	4	5
y	−5	0	3	4	3		

88

Number Puzzle

The difference between two numbers is 12. The product of the same two numbers is 45.

1. Find the two numbers using a symbolic strategy.

2. Find the two numbers using a graphing strategy.

3. Find the two numbers using a table.

4. Is there another way to find the two numbers?

5. Discuss the advantages and disadvantages of each strategy for this situation.

Mr. Wiley's Baby

Mr. Wiley presented some data to his algebra class about the early growth of his new baby boy. Hermione created a scatter plot for Mr. Wiley's data. The data and scatter plot are shown at right.

Week	Weight
1	8.5
2	9.25
3	9.75
4	9.75
5	10.5
6	11
7	11.5
8	11.75

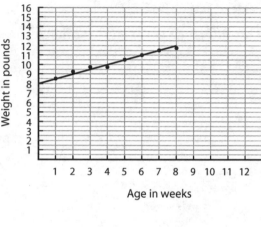

1. Does the data seem to represent a linear model? What equation fits Hermione's graph model?

2. How long do you suppose this growth pattern will continue in this way? If it does continue, what would you expect Mr. Wiley's son to weigh in 1 year? In 2 years? In 14 years?

3. Within what age limits do you think this is a reasonable representation of the growth of Mr. Wiley's baby?

90

Sketch a Graph

Sketch a graph that represents each situation below. Identify your variables and label the axes of your graph accordingly.

1. Edgar is paddling his kayak down the reservoir to visit a friend. For half an hour, he paddled slowly, enjoying the morning. Then he noticed that the weather might be changing and increased his speed. After another 10 minutes, the wind picked up. For about 20 minutes, the wind blew in the direction he was paddling. Then it shifted and blew directly into his face for the last 15 minutes.

2. Jennifer has been taking tennis lessons for 3 months. She learned slowly at first, but her skill has been steadily increasing. Now she is going to visit her cousins in the country and won't be able to play or practice for two weeks.

91

Analyzing Scatter Plots

Ashley is reviewing the results of data collected and graphed from three different experiments. Help her by describing the patterns you see in these graphs. If appropriate, draw lines that you think might fit the data. If you don't think that a line fits the data, explain why.

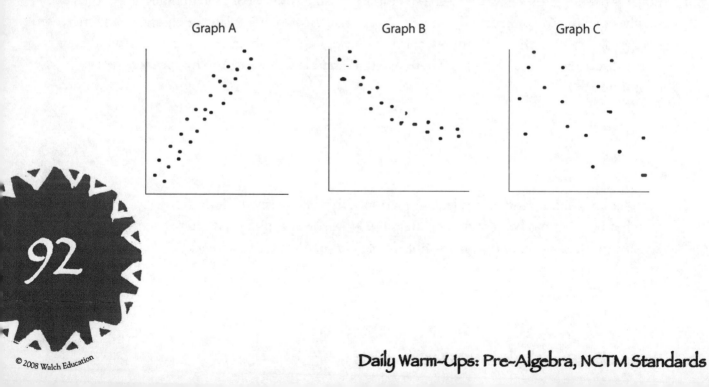

Graph A   Graph B   Graph C

Identifying Graphs from Equations

Equations can reveal a lot of information. You can tell a lot just by how an equation looks and the variables and operations that are in it. Based on what you know about the nature of equations, identify whether each equation below represents a quadratic, a linear, or an exponential relationship. Explain how you made each decision.

1. $y = 7x + x^2$

2. $y = 3x + 7$

3. $y = (6 - x)x$

4. $y = 3^x$

5. $y = 3x(x + 4)$

93

Table Patterns II

Look at the tables below. For each table, do the following:

a. Describe symbolically or in words the pattern hiding in the table.

b. Give the missing values in the table.

c. Tell whether the relationship between x and y in the table represents a quadratic, a linear, an inverse, an exponential, or other relationship. Explain how you know.

1.

x	−1	0	1	2	3	4	5
y	5	7	9	11	13		

2.

x	0	1	2	3	4	5
y	1	2	5	10		

94

Part 4: Understanding Algebraic Symbols

National Council of Teachers of Mathematics: "Instructional programs from pre-kindergarten through grade 12 should enable all students to represent and analyze mathematical situations and structures using algebraic symbols."

Expectations

- Develop an initial conceptual understanding of different uses of variables.
- Explore relationships between symbolic expressions and graphs of lines, paying particular attention to the meaning of intercept and slope.
- Use symbolic algebra to represent situations and to solve problems, especially those that involve linear relationships.
- Recognize and generate equivalent forms for simple algebraic expressions and solve linear equations.

Hedwig's Hexagons

Hedwig is using toothpicks to build hexagon patterns. The first three are pictured. How do you think her pattern will continue? She has begun to collect information in a table in order to explore the relationships among the number of toothpicks, the number of hexagons, and the length of the outside perimeter of the figure. Complete the table for her.

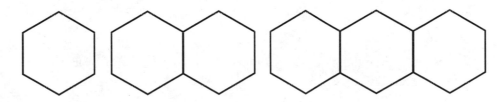

Number of toothpicks	6		
Number of hexagons	1	2	
Perimeter of figure	6	10	

Now write an equation for the relationship between the number of hexagons in each figure and the perimeter of the figure. Write another equation for the relationship between the number of hexagons and the number of toothpicks.

95

Evaluating Simple Expressions I

Evaluate each expression below for the given value of x.

1. $4.5x - 12$ for $x = 2$

2. $-5 - x$ for $x = \dfrac{1}{2}$

3. $2x^2$ for $x = 8$

4. $x^2 + 2.5$ for $x = 1.5$

5. $(x - 2.5)(x + 35)$ for $x = 2.5$

6. $x(27 - x)$ for $x = 3$

7. $\dfrac{49}{x^2}$ for $x = -7$

8. $5x^2 + 3x - 10$ for $x = 2$

9. $\dfrac{x}{4}$ for $x = 8$

10. $.5x^2 + 2x - 20$ for $x = 10$

96

Evaluating Simple Expressions II

Evaluate each expression below for the given value of x.

1. $42 - 3x$ for $x = 6$

2. $13 - 3x^2$ for $x = 1$

3. $4x^2 + 13$ for $x = -10$

4. $6x^2 + x - 2$ for $x = 2$

5. $(x - 2.5)(x + 5)$ for $x = 0$

6. $x(27 - x)$ for $x = -3$

7. $\dfrac{3(15 - x)}{2x}$ for $x = 10$

8. $8x - 3x(6 - x)$ for $x = 0$

9. $\dfrac{x^2}{4}(x + 8)$ for $x = -8$

10. $-2x^2 + 5x + 12$ for $x = -4$

97

Thinking Around the Box

The rectangular box shown below has a front and back length of L meters. The side lengths are W meters. The height is H meters.

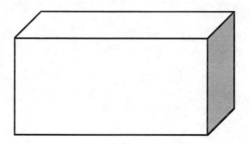

1. Write two equations that represent the sum, S, of all the edges of the rectangular box.

2. Write two equations that represent the surface area, A, of the rectangular box.

98

Perimeter Problem

Use the figure on the right to answer the questions that follow.

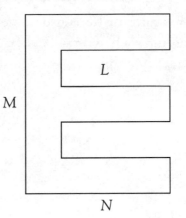

1. If M = 5 meters, N = 4 meters, and L = 2.5 meters, what is the perimeter of the figure?

2. Using the variables only, write two or more expressions that could represent the perimeter of the figure. Show that your expressions are equivalent.

3. Is there another way to represent the perimeter of the figure?

99

Solving Equations 1

Tucker is tutoring his cousin James in pre-algebra. He is trying to explain different strategies for solving an equation such as $53 = 17 - 4x$. What steps or approaches should he recommend to James? Explain your ideas carefully.

100

Solving Equations II

Solve each equation below symbolically. Be prepared to justify your answers.

1. $3x + 7 = 19$

2. $7 + 4x = 31$

3. $26 - 3x = 4x + 19$

4. $x^2 - 3.5x = 0$

5. $13x - x^2 = 0$

6. $x^2 - 10x = 5x^2 - 2x$

7. $4x(x - 5) = 0$

8. $3.5(x + 2) + 2(x + 2) = 0$

9. $5.5x = x - 9$

10. $49x - 7x^2 = 0$

101

Simplifying Expressions

Write at least two more expressions that are equivalent to each given expression below.

1. $8(x - 5)$

2. $x(5x - 6) + 13x - 10$

3. $4(x + 5) - 3(2 - 4x)$

4. $2.5(12 - 2x) + 5(x + 1)$

5. $(3x^2 + 5x + 8) - (8x^2 + 2x - 5)$

102

Slope-Intercept Equations I

Simplify the following equations so that you could enter them more easily into a graphing calculator. Give the slope and y-intercept for each without graphing it.

1. $y - 3(x - 7) = 5x$

2. $y = -2 + (x + 1)$

3. $y + 4 = \dfrac{5}{3}(x + 6)$

4. $y = 28 + 2.5(x - 6)$

5. $y = 13.2(x - 20) + 125.6$

103

Slope-Intercept Equations II

Find an equation written in slope-intercept form that satisfies each condition below.

1. a line whose slope is -3 and passes through the point $(2, 5)$

2. a line whose slope is $\dfrac{-2}{3}$ and passes through the point $(4, -1)$

3. a line that passes through the points $(2, 6)$ and $(6, 1)$

4. a line that passes through the points $(-2, 3)$ and $(4, -3)$

5. a line that passes through the points $(-1, -1)$ and $(-5, -5)$

104

Up and Down the Line I

Write an equation for the line pictured in each graph below.

1.

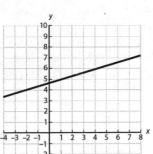

2.

3.

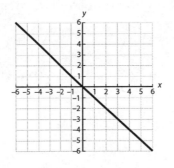

4.

105

Up and Down the Line II

Write a linear equation for each condition below.

1.

x	−1	0	1	2	3
y	1	3	5	7	9

2. a line whose slope is −3 and y-intercept is 5

3. a line that passes through the points (2, 5) and (5, 6)

4. a line that passes through the point (3, 7) and has a slope of $\frac{2}{3}$

Butler Bake Sale

The Butler Middle School eighth grade class is planning a bake sale as a fall fund-raiser. Luis is chairing the planning committee. He gave a brief survey to determine what price should be charged for each brownie. He predicts from his results that a price of $0.50 per brownie will result in 200 brownies sold. He also predicts that a price of $1.00 per brownie will result in about 50 brownies sold. He is assuming that the relationship between the brownie price and the number sold is a linear relationship.

1. Write an equation for the relationship that Luis has predicted between the cost and the number of brownies sold.

2. Find the slope and y-intercept. Then explain what these mean in the context of Luis's information.

3. What if the committee decides to charge $0.70 per brownie? How many can they expect to sell?

4. Luis has taken an inventory and found that the class has 300 brownies to sell. Using your equation, what would be an appropriate price to charge for each brownie?

107

Looking for Lines 1

Explain how you can find the equation of a line if you know the information below. Use examples to explain your thinking.

1. the slope and y-intercept

2. two points on the line

3. the slope of the line and a point that is on the line, but is not the y-intercept

Looking for Lines II

Linear patterns can be recognized in several different representations of a relationship between two sets of quantities. Sometimes we look for the relationship in graphs, tables, or symbolic rules. Describe how you can tell whether a situation is or can be a linear model by looking at the information below.

1. a scatter plot of the data

2. a table of the values

3. the equation of the relationship

4. a description of the problem

Thinking About Equations of Linear Models

Think about ways to find an equation of the form $y = mx + b$ or $y = a + bx$ from a table of data or a graph of the points. How can you find the equation if you know the slope and y-intercept? How can you find the graph by looking for the rate of change and other values from the table? How can you find the equation of the line if the slope and y-intercept are not given? Write a few sentences to explain your thinking.

110

Thinking About Change and Intercepts in Linear Models

Think about linear models in various forms. How can you see or find the rate of change in a linear model from a table of values? How would you find it from the graph? How does it appear in the equation? How would you determine the y-intercept in those three situations? Write a few sentences to explain your thinking.

111

Chirping Crickets

Did you know that you can estimate the temperature outside by counting the number of cricket chirps you hear? One method to estimate the temperature in degrees Fahrenheit is to count the number of chirps in 15 seconds and then add 37. This will give you the approximate temperature outside.

1. Write a mathematical sentence that describes the number of cricket chirps per 15 seconds (C) as a function of the temperature (*T*) in Fahrenheit.

2. Graph this function.

3. What do you think are the maximum and minimum values of this function?

112

Freezing or Boiling?

Lucas knows that the relationship between Celsius and Fahrenheit temperatures is a linear one. However, he often forgets the equation. He remembers that at freezing, the temperatures are (0°C, 32°F). At boiling, the respective temperatures are (100°C, 212°F). Explain to Lucas how he can find this equation knowing that the relationship is a linear one.

The Seesaw Problem

1. Eric and his little sister Amber enjoy playing on the seesaw at the playground. Amber weighs 65 pounds. Eric and Amber balance perfectly when Amber sits about 4 feet from the center and Eric sits about $2\frac{1}{2}$ feet from the center. About how much does Eric weigh?

2. Their little cousin Aleah joins them and sits with Amber. Can Eric balance the seesaw with both Amber and Aleah on one side, if Aleah weighs about the same as Amber? If so, where should he sit? If not, why not?

114

Weekly Pay Representations

Amanda has a part-time job working for a local company selling wireless phone subscriptions. She is paid $120 per week plus $25 for each subscription that she sells.

1. Complete the table below, showing Amanda's weekly pay as a function of the number of subscriptions that she sells.

2. Using variables, write two rules (equations), one for the weekly pay in terms of her sales, and one for her sales in terms of her pay.

3. Amanda wants to earn at least $600 next week. How many subscriptions must she sell?

Symbol Sense: Original Costs

Think of a real-world situation that might result in each symbolic sentence below. Solve the sentences for C.

1. C − .25C = $37.49

2. $28.00 + .20C = C

116

Babysitting Bonus

This year, Zachary has been babysitting his young cousins after school for $70 a month. His uncle also gave him an extra bonus of $100 for his excellent work. Since school started, Zachary has earned more than $500. How many months ago did school start? Write an inequality that represents this situation. Solve it showing all your work.

117

Exponential Expressions

The following exponential expressions can be rewritten in other forms. Use your understanding of how exponents behave to change each expression to an equivalent one.

1. x^{-n}

2. $\dfrac{y^r}{y^s}$

3. $(x^a)^b$

4. $(ab)^n$

5. $\left(\dfrac{d}{t}\right)^n$

6. $\dfrac{1}{x^n}$

7. $n^p \bullet n^q$

8. $y^{\frac{1}{n}}$

118

Nathan's Number Puzzles I

Nathan likes to make up puzzles about integers. Some of his recent puzzles are below. Write symbolic sentences that represent Nathan's puzzles. Then solve each puzzle.

1. Two numbers have a sum of 10. If you add the first number to twice the second number, the result is 8. What are the numbers?

2. One number is twice as large as a second number. The sum of the two numbers is 15. What are the numbers?

3. The first number minus the second number is 2. Twice the first number minus twice the second number is 4. What are the numbers?

119

Daily Warm-Ups: Pre-Algebra, NCTM Standards

Nathan's Number Puzzles II

Nathan has written pairs of linear sentences in symbols. Now he's trying to create word puzzles to go with them. Help him by creating a word puzzle for each pair of linear sentences below. Then find the numbers that fit his equations.

1. $x + 3y = 5$
   $3x + y = 7$

2. $x + 3y = 10$
   $3x + 14 = 2y$

3. $x + 2y = 7$
   $x + 2y = 17$

Quincy's Quiz

Quincy has been confused about simplifying algebraic expressions. His brother Eli prepared this quiz for him to practice. Look at Quincy's responses below. If the response is correct, write *correct*. If it is incorrect, write the correct answer.

1. $4y - 5y$ $-y$

2. $3x + 2x$ $5x^2$

3. $2n^2 + 4n^2$ $6n^4$

4. $2x \bullet 3y$ $5xy$

5. $2a - (a - b)$ $a - b$

6. $\dfrac{n+5}{n}$ 5

7. $4 - (n + 7)$ $-3 - n$

8. $\dfrac{2x^6}{x^3}$ $2x^2$

9. $(2xy)^2$ $4x^2y^2$

10. $2a + 3b + 3a + 4b$ $6a^2 + 12b^2$

121

Daily Warm-Ups: Pre-Algebra, NCTM Standards

Systems of Linear Equations I

Marika has asked you to help her understand how to solve systems of equations. Solve each system of equations below using a different strategy. Then explain to Marika why you chose that strategy for that system. Which are best solved by substitution? Which might be easily graphed? Which could be solved by elimination?

1. $y = x - 1$
    $3x - 4y = 8$

2. $3x + 2y = -10$
    $2x + 3y = 0$

3. $x + y = -10$
    $.5x + 1.5y = 5$

4. $3x - 2y = 6$
    $-2x + 3y = 0$

122

Systems of Linear Equations II

Amal is learning about systems of linear equations. He has come up with some questions regarding equations such as $y = 3x + 8$ and $y = -5x + 11$. Help him by answering the questions below.

1. What is the objective for solving a system of equations such as the one given?

2. How could you find the solution by graphing?

3. How could you find the solution by using tables of values?

4. How might you find the solution for the system without a graph or a table?

5. What would you look for in the table, the graph, or in the equations themselves that would indicate that a system of equations has no solution?

Part 5: Developing Algebraic Thinking Through and for Mathematical Modeling

National Council of Teachers of Mathematics: "Instructional programs from pre-kindergarten through grade 12 should enable all students to use mathematical models to represent and understand quantitative relationships."

Expectations

- Model and solve contextualized problems using various representations, such as graphs, tables, and equations.

Patricia's Patterns

Patricia has been exploring number patterns consisting of 4 consecutive integers. She has noticed a relationship between the products of the inside numbers and the outside numbers. Complete Patricia's table and then tell what you think Patricia's discovery is. Will this always be true? Justify your answer algebraically.

Number	Inside product	Outside product
3, 4, 5, 6		
6, 7, 8, 9		
10, 11, 12, 13		
15, 16, 17, 18		
40, 41, 42, 43		

Daily Warm-Ups: Pre-Algebra, NCTM Standards

Stacking Rods I

Cuisenaire rods are wooden materials often used for mathematics activities in elementary classrooms. They were invented by Georges Cuisenaire in 1952. They consist of 10 rectangular rods with height = 1 cm, width = 1 cm, and lengths of 1–10 cm. Most Cuisenaire rods follow the system below:

white rod = 1 cm
red rod = 2 cm
light green rod = 3 cm
purple rod = 4 cm
yellow rod = 5 cm
dark green rod = 6 cm
black rod = 7 cm
brown rod = 8 cm
blue rod = 9 cm
orange rod = 10 cm

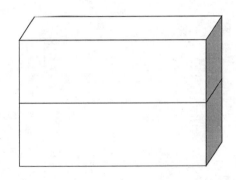

Pierre is making stacks of purple rods and recording the surface area of the exposed sides in a table. What is the surface area of a stack of n purple rods? (Assume that the sides of the rods are exposed, but not the bottom surface.)

125

Stacking Rods II

Marie is investigating purple Cuisenaire rods that are stacked in a staircase.

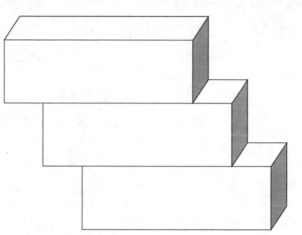

The rods are displaced in each step by exactly 1 unit. Remember that the purple rod is 4 units long. Find the surface areas of stacks of different numbers of rods. Then find a rule or a formula for finding the surface area of a staircase of *n* number of rods. (Assume that the bottom face of each rod is unexposed.)

126

Water Balloon Bungee Jump

Natalie and Celeste are conducting an experiment using rubber bands and water balloons. Their challenge is to predict the distance a water balloon will fall for any number of rubber bands that they might use. This is the data that they have collected so far:

Number of rubber bands	3	4	5	6	7	8	9	10	11
Distance fallen (cm)	71	83	95	115	128	144	160	175	190

They plan to test their conclusions from this data by dropping a water balloon from a stairwell at a starting height of 12.5 feet. How many rubber bands should they use to get the water balloon as close to the floor as possible, but not break it?

127

How Many Handshakes?

It is a tradition in the Magnolia Magnet School of Meticulous Manners for all students and teachers in a classroom to shake hands with every other person in the room. On a typical Monday morning, there are 16 people in Mr. Manners' classroom. How many handshakes will there be? Find a way to determine how many handshakes there will be in any classroom on any day.

128

Spring Break Flight Costs

Ray Airhart is chief pilot for Eagle Valley Aviation. He is promoting a charter flight to Florida for the eighth grade class spring break trip. The flight will cost $200 per person for 10 students or fewer. For every 4 additional people who go on the trip, he can lower the price for each person an additional $16. How many people will make the most money for Eagle Valley Aviation?

Pet Mice

Linn has been keeping several mice as pets for some time. She has recently learned that she can raise mice to sell to pet stores. However, she would need to provide the mice in lots of at least 200. To predict when she will have enough mice to sell, she has collected the data in the chart below.

Month	Mice
1	3
2	7
3	17
4	36
5	65

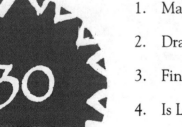

1. Make a scatter plot of Linn's data.

2. Draw a line of best fit.

3. Find an equation in slope-intercept form for your line.

4. Is Linn correct in using a linear model to predict when she will have 200 mice?

5. If so, predict when. If not, explain why not.

Relating Temperature and Altitude

Keri has just begun pilot lessons. She is very interested in the relationships among altitude, pressure, temperature, and density. She did some online research and found the table below. It gives the standard values of pressure, temperature, and density at different altitudes.

Altitude (ft)	Pressure (Hg)	Temperature (°F)	Density (%)
sea level	29.92	59.0	100
2,000	27.82	51.9	94.3
4,000	25.84	44.7	88.8
6,000	23.98	37.6	83.6
8,000	22.22	30.5	78.6
10,000	20.57	23.3	73.8
12,000	19.02	16.2	69.3
14,000	17.57	9.1	65.0
16,000	16.21	1.9	60.9

Use a graphing calculator to make a graph of the relationship between altitude and temperature for $0 \leq A \leq 20,000$. Find a regression equation that you think works for this data. Be ready to explain your thinking.

131

Daily Warm-Ups: Pre-Algebra, NCTM Standards

Fencing Fernando

Franny lives on a small farm with many unusual animals. She has just been given a Vietnamese potbellied pig named Fernando. But before she can get Fernando, she needs to build a pen for him. If needed, she can use one side of a long barn that measures 200 feet. She also knows that she has 120 feet of fencing available. Answer the questions below about Fernando's pen. Be prepared with drawings and mathematical models to justify your answers.

1. What are the longest sides that a rectangular pen could have if the barn is used for one side?

2. Suppose Franny wants a square pen. What are the dimensions of the square, if all the fencing and one side of the barn is used?

3. What is the maximum area that Franny can create for her pig?

4. What is the minimum area that Franny can create?

5. Suppose Franny decides to build her pen in a field, using all of the fencing but not the side of the barn. What are the maximum and minimum areas of her pen?

132

Boxes, Little Boxes

Jaden is exploring the characteristics of small boxes. He is creating boxes using one sheet of construction paper for each box. He makes the box by cutting out squares in 1-inch increments from the four corners of the paper and then folding up the edges. If the paper that Jaden is using is 9 inches by 12 inches, what are the sizes of the boxes that he is designing? Create a table of values that records the length, width, and volume of all the boxes you think Jaden could create in this way. Also include the length of the square cutouts, the surface area of the box, and any other data of interest to you. Do you see patterns in your data? What is the length of the square cutouts for the box with the greatest volume? With the smallest volume? If s represents the side of the square that Jaden cuts, how could he represent the volume of the resulting box in terms of s?

133

Working for Uncle Daniel

Your Uncle Daniel has proposed an unusual payment scheme for a summer job that you are planning to do for him. He says that he will pay you $1,000 in the morning on the first day, but at the end of the day you must pay him a commission of $100. The next day he will double what you cleared the first day, but you must double what you pay him for commission. This arrangement will go on for as long as you wish to work for him. Create a table, a graph, and an argument to present to your uncle that explains and justifies your decision of how long you will work for him.

134

The Growing Cube

A cube is having its first birthday. This cube is called a *unit cube*. Describe how many corners (vertices), faces, and edges it has. On the cube's second birthday, its edges are twice as long as on its first birthday. How many unit cubes would be needed to equal the volume of this new 2-year-old cube? On the cube's third birthday, its edges are 3 times as long as those of the unit cube. This pattern continues for each birthday. How many unit cubes will there be for its third, fourth, fifth, sixth, seventh, eighth, ninth, and tenth birthdays? Write a rule to calculate the number of unit cubes needed for any number of birthdays. Then draw a graph of the number of unit cubes as a function of the number of years old the cube is.

135

Coloring Cubes

Crystal and T.J. are considering cubes of many sizes that have been dipped completely in paint. They start by thinking about a cube that is 10 units long on each side. After it is dipped and dries, the 10-unit cube is cut into single-unit cubes.

1. How many of these newly cut unit cubes are painted on 3 faces, 2 faces, 1 face, and zero faces? Answer those same questions for cubes that are 2, 3, 4, 5, 6, 7, 8, and 9 units long on each side. Then make a chart of your findings for all the cubes. Look for patterns.

2. Where are the unit cubes with 3 painted faces located on the various cubes? With 2 painted faces? With 1 painted face? With zero painted faces?

3. Make a prediction for a cube that is 15 units long on one side. Describe the patterns using symbolic rules that could be used for any size cube.

136

Daily Warm-Ups: Pre-Algebra, NCTM Standards

Credit-Card Charges

Charles purchased a new flat-screen television for $2,000 with his credit card on July 1. His credit-card company charges 1.5% interest on the unpaid balance.

1. If Charles does not pay the credit-card company at the end of one month, how much will he owe?

2. How much will he owe at the end of two months?

3. How much will he owe at the end of three months?

4. When will his unpaid balance reach $2,500?

137

Tyrone's Toothpick Triangles

Tyrone is creating patterns with toothpicks.

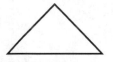 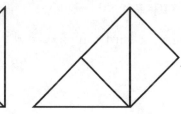 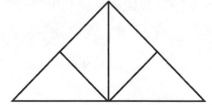

| first figure | second figure | third figure | fourth figure |
| 3 toothpicks | 5 toothpicks | 7 toothpicks | 9 toothpicks |

1. How many toothpicks are in the eleventh figure in his pattern?

2. Write both an explicit equation and a recursive equation for the number of toothpicks, t_n, for the nth figure in Tyrone's pattern of toothpicks.

3. What would the figure look like in Tyrone's pattern that has exactly 71 toothpicks?

138

Switching Chips

Rocco is exploring a children's puzzle that involves a strip of 7 squares, 3 white chips, and 3 black chips.

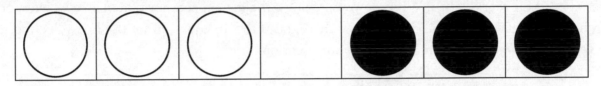

The object of the puzzle is to switch positions of the chips so that all the black chips are on the left and all the white chips are on the right. A chip can move by sliding to an adjacent empty square or by jumping over one chip to land on an empty square. White chips only move right, and black chips only move left.

1. How many moves will it take to switch the chips?

2. Complete the table to show the number of moves it would take for other numbers of chips.

Number of each color	1	2	3	4	5	6	7	8	9	10
Number of moves										

3. A puzzle with n white chips and n black chips will require how many moves to complete the switch?

139

Life Expectancy

The table below shows the average number of years a person born in the United States can expect to live based on their year of birth. What do you predict is the life expectancy of a man, a woman, or any person born in 2020? Find a line of best fit for this data. Write an equation for the line you choose. Make your prediction using your graph and your equation.

Year	Female	Male	Both sexes
1930	61.6	58.1	59.7
1940	65.2	60.8	62.9
1950	71.1	65.6	68.2
1960	73.1	66.6	69.7
1970	74.7	67.1	70.8
1975	76.6	68.8	72.6
1980	77.4	70.0	73.7
1985	78.2	71.1	74.7
1990	78.8	71.8	75.4
1995	78.9	72.5	75.8
2000	79.7	74.3	77.0
2005	80.4	75.2	77.9

140

Interpreting Graphs 1

Which graph best represents the given situation? Be prepared to justify your answers.

1. A commuter train pulls into a station and drops off all its passengers.

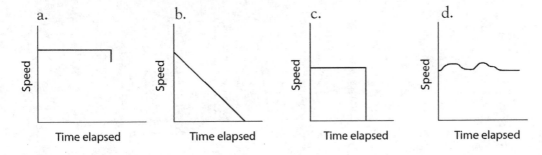

a.   b.   c.   d.

2. A child walks to a slide, climbs up to the top, and then slides down.

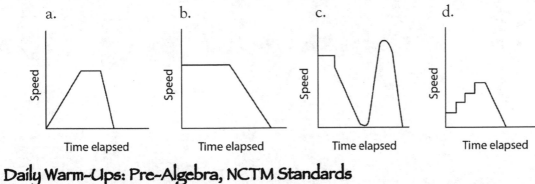

a.   b.   c.   d.

Interpreting Graphs II

Which graph best represents the given situation? Be prepared to justify your answers.

1. A little boy swings on a swing at a playground.

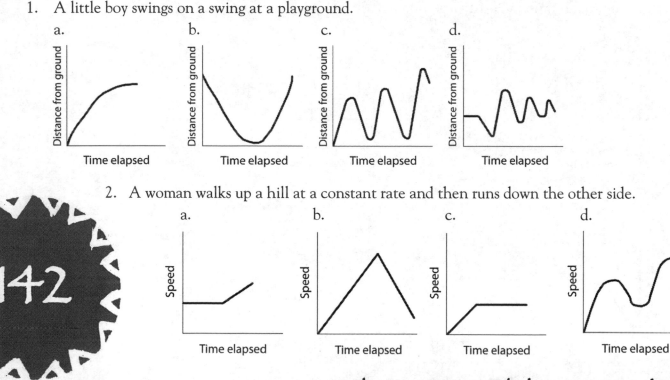

a.

Distance from ground
Time elapsed

b.

Distance from ground
Time elapsed

c.

Distance from ground
Time elapsed

d.

Distance from ground
Time elapsed

2. A woman walks up a hill at a constant rate and then runs down the other side.

a.

Speed
Time elapsed

b.

Speed
Time elapsed

c.

Speed
Time elapsed

d.

Speed
Time elapsed

Daily Warm-Ups: Pre-Algebra, NCTM Standards

Ferris Wheel Graphs

Enrique took a ride on a Ferris wheel at the amusement park last weekend. He has sketched these graphs to represent that ride. Which one best represents the idea that he's trying to show? Explain why you chose that graph.

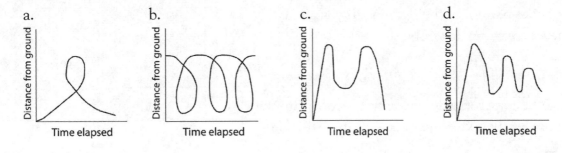

a.

b.

c.

d.

143

Roller Coaster Ride

Marjorie and Jack love to ride roller coasters and then portray their ride in a graph. Last Saturday, they rode the Mighty Twister and sketched this graph of that experience.

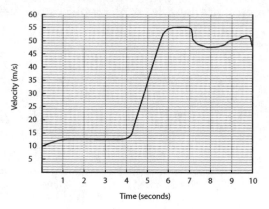

1. What does the pattern of their graph tell you about when they were going down the steepest hill?

2. What was their approximate velocity 1 second into the ride? What was their velocity 8 seconds into the ride?

3. After getting started, when were they going the slowest? The fastest?

4. Describe what was happening to them between 4 and 8 seconds.

U.S. Population

The graph below shows the United States population from 1900 to 2000, as recorded by the U.S. Census Bureau. Use the graph to answer the following questions.

1. What was the rate of change in population from 1900 to 2000? Is this greater or less than the rate of change in the population from 1990 to 2000?

2. Which 10-year time periods have the highest and the lowest rates of change? How did you find these? What factors might have contributed to these rates of change?

3. What do you predict the U.S. population will be in 2010? Explain your reasoning.

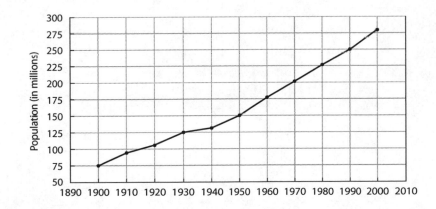

Daily Warm-Ups: Pre-Algebra, NCTM Standards

Phone Plan Decisions

Aisha's parents are planning to give her a new cell phone for her upcoming birthday and pay for her service for one year. They have asked her to determine which plan is best for her. Best Talk offers service at a monthly basic fee of $20.00 per month and $0.10 per minute for each minute used. Horizon One-Rate has no monthly fee, but charges $0.40 per minute. Create a table, a graph, and equations that model this situation. Then give an argument for why Aisha might choose either plan.

146

Fair Trade

Land developers sometimes make deals with landowners and trade pieces of land that they need for other plots of land in other places. Suppose you own a square piece of property that the ABC Land Development Corporation would like to have to build a new mall. The company is willing to trade your property for a rectangular piece nearby that is 3 meters longer on one side and 3 meters shorter on the other side. How does this rectangular piece compare to your original square plot? Would this be a fair trade for any side length of your original square? Complete the table and compare the areas. Write expressions for the situation.

| Original square | | Rectangular plot | | | Difference in area |
Side length	Area	Length	Width	Area	
4	16	7	1	7	9
5					
6					
7					
8					

147

Used Car Purchase

Mary Ellen has just started a new job and will need a car to drive to work. She stopped by Jolly Joe's dealership and found a car that she really likes. Joe has told her that she can buy the car for a down payment of $2,000 and a monthly payment of $399 for 24 months.

Happy Hank runs a competing dealership nearby. He has a car almost exactly like the one Mary Ellen wants. Hank will sell her the car for a down payment of $1,600 and a monthly payment of $420 for 24 months.

1. Write an equation for each situation that represents the amount that Mary Ellen will have paid up to that point for any month in the 24-month period.

2. What amount will she have paid at the end of 1 year if she buys Jolly Joe's car? What if she buys Happy Hank's car? What will she have paid for each car at the end of 2 years?

3. Which is the better deal?

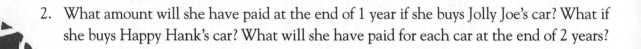

148

The Shape-Shifting Square

A square has a side of length s. A new rectangle is created by increasing one dimension by 5 centimeters and by decreasing the other dimension by 4 centimeters.

1. Draw a sketch that represents this situation.

2. Write symbolic expressions that represent the areas of the two figures.

3. For what values of s will the area of the new rectangle be greater than the original square?

4. For what values of s will the area of the new rectangle be less than the original square?

5. For what values of s will the areas of the two figures be equal?

6. Explain how you found your answers.

Part 6: Analyzing Change in Various Contexts

National Council of Teachers of Mathematics: "Instructional programs from pre-kindergarten through grade 12 should enable all students to analyze change in various contexts."

Expectations

- Use graphs to analyze the nature of changes in quantities in linear relationships.

Parachuting Down

Cecelia took her first parachute jump lesson last weekend. Her instructor gave her the graph below that shows her change in altitude in meters during a 2-second interval. Use the graph to answer the questions that follow.

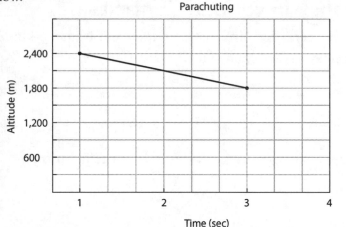

1. What is the slope of the line segment?

2. Estimate Cecelia's average rate of change in altitude in meters per second.

3. Give the domain and range for this graph.

150

Ping-Pong Prices

Lily found the price of ping-pong balls listed on the Internet at $4.75 for a package of 6 balls. Shipping and handling was listed at $1.00 per package.

1. Write an equation that represents the total cost for different numbers of packages of ping-pong balls.

2. Sketch a graph of this relationship.

3. If you shift your graph up a value of $.50, does this mean the price per package increased, or the shipping price increased?

4. Write a new equation for the situation in question 3.

151

Daily Warm-Ups: Pre-Algebra, NCTM Standards

Graphing People Over Time

Think about each situation below. Then sketch a graph to represent the situation over a 24-hour period of time. Label each graph carefully using the horizontal axis to represent the time of day. Be prepared to explain and justify your choices.

1. the number of people at a very popular pizza restaurant on a Saturday

2. the number of people in a school building on a weekday in September

3. the number of people at a sports stadium on the day of a big game

4. the number of people at a movie theater on a weekend day

152

DVD Rentals

Rosa's parents just bought a new DVD player for the family. Rosa's mom asked her to research rental prices for DVDs online. More Movies has a yearly membership package. Rosa found the following table of prices on the company's web site.

More Movies DVD Rental-Membership Packages

Number of videos rented	0	5	10	15	20	25	30
Total cost	$30	$35	$40	$45	$50	$55	$60

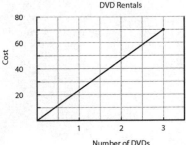

On the Deluxe DVD Rentals web site, Rosa found that there was no membership package offered. She made the graph on the right to show how the cost at Deluxe DVD Rentals is related to the number of videos rented.

If both rental services have a comparable selection of DVDs and Rosa's family will watch on average about two movies a month, which service should they choose? Explain how Rosa's family might decide which service to use. Using the information about each service, describe the pattern of change relating the number of DVDs rented to the total cost.

Camping Costs

Dan and Julia are planning a camping trip for their school's Outdoor Adventure Club. They went online and found the following information for Hamden Hills State Park. Use the information to answer the questions below.

Number of campsites needed	1	2	3	4	5	6	7
Total campground fee	$17.50	$35.00	$52.50	$70.00	$87.50	$105.00	$122.50

1. Make a coordinate graph of these data. Would it make sense to connect the points on the graph? Why or why not?

2. Using the table, describe the pattern of change in the total fee as the number of campsites needed increases. How is this pattern shown in your graph?

154

Daily Warm-Ups: Pre-Algebra, NCTM Standards

Renting Canoes

Dan and Julia have reservations for 10 students for the Outdoor Adventure Club's annual trip. This year they are planning a canoe camping trip. The Otter Mountain River Livery rents canoe and camping gear for $19 per person. Dan and Julia expect no more than 50 students in all to go on the trip. Using increments of 10 campers, make a table showing the total rental charge for 10 to 50 campers. Then make a coordinate graph of these data.

155

Counting Step-Ups

Alejandro has collected data on the number of step-ups his friend Jalesia can do in 2 minutes. The graph shows their data.

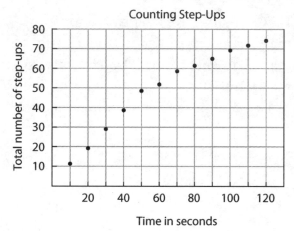

1. What are the two variables represented on the graph?

2. Make a table of Alejandro's data. Describe how the number of step-ups changes for each 10-second interval as time increases.

3. From Alejandro's data, estimate the number of step-ups Jalesia did in 25 seconds and in 65 seconds. Explain how you made those estimates.

156

Grow Baby!

The average growth weight in pounds of a baby born in the United States is a function of the baby's age in months. Look at the sample data provided and graph the information in an appropriate window. Then write two sentences about the graph and the relationship between the two variables.

Age in months	Weight in pounds
0	7
3	13
6	17
9	20
12	22
15	24
18	25
21	26
24	27

157

Braking/Stopping Distances

Use a graphing calculator or graph paper to graph the relationship between miles per hour and stopping distance using the data from the table below. What questions come to mind as you look at your graph? Make a prediction for the stopping distance at 80 miles per hour. Explain and justify your prediction.

Miles per hour	Stopping distance
10	27
15	44
20	63
25	85
30	109
35	136
40	164
45	196
50	229
55	265
60	304
65	345

158

Daily Warm-Ups: Pre-Algebra, NCTM Standards

Thinking About Variables, Graphs, and Tables

1. Think about a situation in which variable y depends on variable x. (For example, y might be profit and x the number of items sold.) If y increases as x increases, how would this appear in a table? How would it appear in a graph?

2. Now think about a situation in which variable y decreases as variable x increases. (For example, y might be the amount of gasoline in your car on a trip and x the time you have been traveling.) How would this be indicated in a table? In a graph?

3. In a coordinate graph of two related variables, when do the points lie in a straight line?

4. In a coordinate graph of two related variables, when is it appropriate to connect the points?

159

Percent Increase or Decrease

Look at the sequences below. For each, tell whether there is a growth or decay, identify the common ratio, and give the percent increase or decrease.

1. 43, 129, 387, 1,161, . . .

2. 90, 99, 108.9, 119.79, . . .

3. 1,800, 1,080, 648, 388.8, . . .

4. 17.8, 3.56, 0.712, 0.1424, . . .

5. 375, 142.5, 54.15, 20.577, . . .

Intervals of Change

Look at the table and graph on the right.

x	y
−5	−10
−4	0
−2	13
0	16
2	14
4	2
5	−5

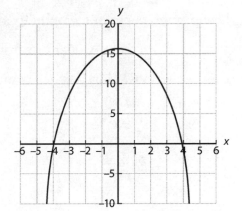

1. Identify the x- and y-intercepts, if any. How is the table helpful? How is the graph useful?

2. Consider the x interval (−5, 0). Describe how y may be changing over that interval of x.

3. Now consider the x intervals (0, 2), (2, 4), and (4, 5) for this graph and table. Is the average rate of change constant over these intervals?

Looking at Graphs

Using graph paper or a graphing calculator and a window of 0 to 15 on each axis, show the graphs of $y = \dfrac{8}{x}$ and $y = 8x$ for values of x from 1 to 10. What are the least and greatest values for y in each of these equations? How do they compare? Where do the two graphs intersect?

162

Rolling Along

Did you know that you can calculate the speed (S) of a car by knowing the size of its tires and the revolutions per minute (RPM) that the tires make as they rotate? Think about a car that has tires that are 2.5 meters in circumference. Imagine that the tires are rotating at 500 revolutions per minute.

1. What is the speed of the car?

2. Write verbal and symbolic rules that express the relation between time and distance.

3. Create a table that records some specific data pairs (time, distance) of your choice.

4. What would be the distance traveled in 20 minutes?

5. Sketch a graph that represents the data you have in your table in number 3. Graph the independent variable on the horizontal axis and the dependent variable on the vertical axis.

163

Looking for Difference Patterns

Simone has been studying the graphs and characteristics of quadratic equations (those equations that have an x^2 in them). She has been looking at the equations below. She wants to create a table like the one below for each of them. Finish the table and then make a new table for each additional equation using integer values for x between -5 and 5.

1. $y = 3x \, (x + 2)$

2. $y = 2x - x^2$

3. $y = x^2 + 5x + 6$

x	$y = 3x \, (x + 2)$	First difference in y's	Second difference
-5	45	$45 - 24 = 21$	$21 - 15 = 6$
-4	24	$24 - 9 = 15$	$15 - 9 = 6$
-3	9	$9 - 0 = 9$	
-2	0		
-1			
0			
1			
2			
3			
4			
5			

Describe any patterns that you see in the first-difference and the second-difference columns for each equation. How are these patterns different, and how are they alike? What do the graphs of these equations look like?

164

Daily Warm-Ups: Pre-Algebra, NCTM Standards

Part 7: Data Analysis and Probability

National Council of Teachers of Mathematics: "Instructional programs from pre-kindergarten through grade 12 should enable all students to formulate questions that can be addressed with data and collect, organize, and display relevant data to answer them, select and use appropriate statistical methods to analyze data, develop and evaluate inferences and predictions that are based on data, and understand and apply basic concepts of probability."

Expectations

- Select, create, and use appropriate graphical representations of data, including histograms, box plots, and scatter plots.

- Find, use, and interpret measures of center and spread, including mean and interquartile range.

- Discuss and understand the correspondence between data sets and their graphical representations, especially histograms, stem-and-leaf plots, box plots, and scatter plots.

- Make conjectures about possible relationships between two characteristics of a sample on the basis of scatterplots of the data and approximate lines of fit.

- Use proportionality and a basic understanding of probability to make and test conjectures about the results of experiments and simulations.

- Compute probabilities for simple compound events, using such methods as organized lists, tree diagrams, and area models.

Rolling Dice

1. Think about rolling a standard pair of fair dice. What are all the possible sums that you could get from the two dice?

2. What is the probability of rolling each sum?

3. How does the probability of rolling a sum of 7 or 11 compare to the probability of NOT rolling a 7 or 11 on your first roll of two dice? Explain your thinking.

165

Pairing Socks

Jamal does not like to take the time to pair his socks after he takes them out of the dryer, so he just throws them into his sock drawer. He knows he has 4 black socks, 2 blue socks, 2 white socks, and 2 tan socks in his sock drawer. One morning he needs to dress in a hurry and wants to wear his tan socks. Without turning on his bedroom light, he reaches in and pulls out one sock.

1. What is the probability that the sock he pulls out is tan?

2. What is the probability that he will get a black or a blue sock on the first try?

3. If he pulls out a tan sock the first time, what is the probability that he will get a tan sock on his next try?

4. What is the probability that Jamal will get a matching pair of socks of any color if he makes two successive selections?

166

Tossing Coins

1. If you toss a nickel and a dime together, what is the sample space (or all the possible results that you could get)?

2. What is the probability that you would get heads for the nickel and tails for the dime on the same toss?

3. What is the probability that one coin will be heads and one will be tails?

4. What is the probability that *at least* one coin will be tails?

167

Ice-Cream Cones

Frosty Freeze features 9 different ice-cream flavors each Wednesday. How many different 2-scoop cones could you order if you did not repeat flavors? What if you allowed for 2, scoops of the same flavor as well? Justify your thinking. What if the order of scoops matters? What if order does not matter?

168

Spinning for Numbers

Look at the spinner below. If it is spun, find each probability given.

1. P(a factor of 12)

2. P(a multiple of 3)

3. $P(9)$

4. P(a prime number)

5. P(an even number)

6. P(neither a prime nor a composite number)

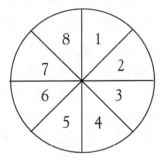

169

Fair Game

In a fair game, each player has an equal chance of winning. Read the description of the game below. Then answer the questions that follow.

An unbiased non-player always spins the spinner shown. There are two contestants, Player A and Player B. For each spin, Player A always multiplies the number spun by 2, and Player B adds 4 to the number spun. The player whose total is greater on that spin gets 1 point. If the two values are equal, no one scores on that spin. For example, a 4 is spun. Player A multiplies 4 times 2 to get 8. Player B adds 4 and 4 to get 8. No one scores for that round. The first player to score 20 points wins.

1. Do you think the winner (Player A or B) would be the same for most games? Write a few sentences to explain your answer. What is the probability of each player winning a point on any given spin?

2. Is the game fair? If not, how could the rules be changed to give each player an equal chance to win without changing the values on the spinner? Make sure to show how the probability of winning on a given spin would be the same for each player in the new game.

Daily Warm-Ups: Pre-Algebra, NCTM Standards

An Unfair Game

Using the spinner pictured below, design rules for a game that would involve two players (Player A and Player B), in which Player A would have twice the chance of winning as Player B.

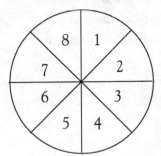

171

Simulating Free Throws

Sami plays on his school's basketball team. His coach told him that his most recent free-throw shooting percentage is 60%. His friend Mariel has designed the spinner below to simulate Sami shooting foul shots. Suppose that Sami was fouled in a ball game and was sent to the foul line to shoot a one-and-one. (One-and-one means that the player must make the first foul shot in order to take a second shot.) What is the probability that Sami will score 0 points, 1 point, or 2 points on this one-and-one foul? How could you use the spinner to find the experimental probability for this situation? Create a tree diagram to investigate the theoretical probability.

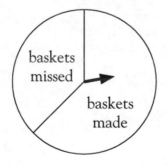

baskets missed

baskets made

172

Comparing the National Debt

1. In 2006, the national debt was $9,207,177,000,000. If the population in 2006 was 304,209,000, how much was the debt per person?

2. If the median household income was $48,201 in 2006, how many households would it have taken to pay off the debt?

3. The highest paid executive in 2006 was Thomas J. Fitzpatrick, CEO of Sallie Mae, a financial services company. He made $39,629,325 in total compensation. How many of Mr. Fitzpatrick's annual incomes would it have taken to pay off the national debt?

173

World Carbon Emissions

The table to the right contains information from the U.S. Energy Information Administration. It gives the carbon-dioxide emissions for fourteen of the more industrialized countries in the world.

Carbon-Dioxide Emissions from Consumption of Fossil Fuels by Country: 1980 to 2004									
Region/Country	1980	1985	1990	1995	2000	2001	2002	2003	2004
	(Million metric tons of carbon equivalent)								
World total	5,000	5,294	5,843	6,009	6,505	6,578	6,668	6,999	7,376
Australia	54	61	72	78	96	100	102	101	105
Brazil	51	51	61	79	94	96	96	87	92
Canada	123	119	131	138	155	151	154	162	160
China	397	501	611	784	827	869	902	1,063	1,284
France	133	108	101	102	109	110	109	111	111
Germany	287	275	189	239	231	237	230	235	235
India	82	120	160	236	273	278	279	284	304
Italy	100	102	113	117	121	120	122	128	132
Japan	256	244	277	293	325	319	323	339	344
Mexico	63	74	82	87	104	103	105	107	105
Netherlands	52	50	56	60	68	75	70	70	73
Poland	115	114	90	83	79	75	74	78	78
United Kingdom	166	160	163	151	150	154	151	154	158
United States	1,297	1,251	1,367	1,443	1,586	1,566	1,570	1,584	1,612

1. What is the range for the emissions for these countries in 2004?

2. What is the difference between the mean for the total emissions in 1980 and the mean in 2004?

3. In 2004, U.S. emissions were what percentage of the world total?

174

Bread Prices

1. According to the U.S. Bureau of Labor Statistics, a loaf of white bread cost $0.70 in 1990 and $0.99 in 2000. What is the percent increase in price for this 10-year period?

2. If the trend continues, what would be the predicted price for a loaf of white bread in 2010?

3. If the price in 2005 was $1.29, is the trend continuing at the same percentage rate? Explain your thinking.

175

Create a Spinner

Jesse and Josh are creating a spinner game. They need a spinner that has 5 regions that are colored red, blue, yellow, green, and purple. They want the regions to have the following probabilities: $P(\text{red}) = \frac{1}{4}$, $P(\text{blue}) = \frac{3}{8}$, $P(\text{yellow}) = \frac{1}{6}$, $P(\text{green}) = \frac{1}{8}$, and $P(\text{purple}) = \frac{1}{12}$. Create the spinner that the boys need for their game.

176

Daily Warm-Ups: Pre-Algebra, NCTM Standards

Coin Chances

1. Alicia tossed a quarter 5 times in a row. It landed tails up 5 times in a row. What is the probability that it will land tails up when she tosses it again? Explain your thinking.

2. Daryl has 5 coins worth exactly 27 cents in his pocket. What is the probability that 1 coin is a quarter? What is the probability that 3 coins are nickels?

3. Make up another probability question about Daryl's coins. Provide the answer, and be ready to explain your thinking.

177

Carnival Dice

The eighth-grade class at Foggy Valley Middle School is planning a fund-raising carnival. Jacob and Austin are members of the games committee. They are designing a game using two huge 6-sided cubes numbered from 1 to 6. A player wins if he or she gets a sum of 7 or 10 on one roll. Jacob and Austin plan to charge $1.00 per roll. Players who win will receive $4.00. Will this be a good fund-raiser for the class? What is the probability of winning? If a player rolls 20 times, how much money could he or she expect to win? Is this a fair game? Write a few sentences to explain your thinking.

178

Ring Toss

Shareen and her brother Samir like to go to carnivals and fairs in the summertime. They enjoy the games and the rides. One day, they watched a ring-toss game in which there were old glass soda bottles standing on a wooden platform. The attendant was encouraging Samir to play the game. He said that Samir had a 50% chance of getting any ring to fall over a bottle because the ring will either go on the bottle or fall off. Do you think that Samir and Shareen should believe the attendant? Why or why not? Explain your thinking.

179

The Probability of Hearts

1. A standard deck of playing cards has 4 suits—diamonds, hearts, clubs, and spades. Each suit has 13 different cards. If you randomly choose a card from a deck, what is the probability that you will get a heart?

2. If you draw 12 cards, how many hearts would you expect to get?

3. If you take all the clubs out of the deck of cards and then draw 12 cards again, how many hearts would you expect to get?

4. Do the answers to the previous questions reflect theoretical probability or experimental probability?

Part 1: Number Sense and Number Theory

1. The fraction 7/8 is larger. Brigit's strategy: 7/8 = .875; 2/3 = .666

   Jordan's strategy:

   Jose's strategy:

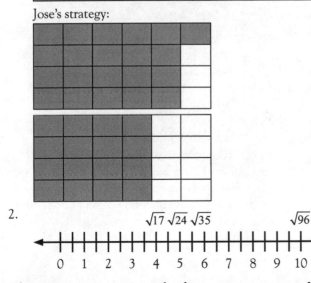

2.

$\sqrt{17}\ \sqrt{24}\ \sqrt{35}$ $\sqrt{96}$

0 1 2 3 4 5 6 7 8 9 10

3. $2^{50} = 1.125899907 \times 10^{15}$ sheets of paper /1,000 = $1.125899907 \times 10^{12}$ inches /12 = $9.382499224 \times 10^{10}$ feet /5,280 = 17,769,884.89 miles

4. Students should give examples of multiplying and dividing with fractions to show that sometimes multiplication results in smaller products and division results in larger quotients than the number originally being multiplied or divided.

5. Veronica can make 9 bows, with $\frac{1}{3}$ yard remaining.

6. Answers will vary. Sample answer: Sam needs $\frac{2}{3}$ yard of fabric to make an apron. If she has 5 yards of fabric, how many aprons can she make? She can make 7 aprons with $\frac{1}{2}$ a yard remaining.

7. Max is right. Possible thinking: Both $\frac{3}{4}$ and $\frac{2}{3}$ are greater than $\frac{1}{2}$, so the result should be greater than 1, but $\frac{5}{7}$ is less than 1. Students also might discuss the common denominator equaling 12.

8. Students could scale to 60 roses, which yields $75 and $69 respectively, or find the price per rose, which yields $1.25 and $1.15 respectively. Thus, 20 roses for $23.00 is the better buy per rose.

9. 3.44 miles per hour

10. There were 301 T-shirts. Students will likely use least common multiples of 4, 5, and 6 that are divisible by 7.

11. 6, 28, and 496 are perfect numbers.

12. Anatole was born on a Saturday (unless a leap year was involved). 759/7 = 108 R3

13. The GCF is 12. Find the prime factorization of each number and compare. Then take the product of the common factors.

14. They will all be there again in 300 days; this number represents the least common multiple (LCM).

15. 1. 221 years
    2. 336 years
    3. The answer represents the LCM.

16. Answers will vary. Sample answers: (12 + 34) + (5 × 6) + 7 + 8 + 9 = 100; 123 − 4 − 5 − 6 − 7 + 8 − 9 = 100

17. Students should get 6 each time.
    $\{[n + (n + 1) + 11]/2\} - n = 6$

18. 1. Rectangles of 36 and 48 square yards will have the most configurations because they have more pairs of factors.
    2. 36 and 49 will have square configurations.

19. There are 1,024 pages: 9 one-digit pages, 90 two-digit pages, 900 three-digit pages, and 25 four-digit pages.

20. Garrett can make 8 different configurations: 1 × 144, 2 × 72, 3 × 48, 4 × 36, 6 × 24, 8 × 18, 9 × 16, and 12 × 12. These represent pairs of factors.

21. $\frac{9 \times 8}{2} = 36$ games

22. 1. $\frac{60}{100} = \frac{6}{10} = \frac{3}{5} = 0.6 = 60\%$
    2. $\frac{55}{100} = \frac{11}{20} = 0.55 = 55\%$

23. $0.3 \times 0.4 = 0.12; \frac{3}{10} \times \frac{4}{10} = \frac{12}{100}$; 30% of 40% is 12%

24. There will be no remainder. The friends predict that they will eat 110% of the pizza. (10% +50% + 35%+15% =110%)

25. They sold 103 cookies at $0.31 each.

26. 45 pies (or 44, depending on rounding)

27. 1. $\frac{7}{10}, \frac{13}{20}$. In addition to equivalent-fraction procedures, students could use fraction strips or fraction circles to show the relationships between the fractions.
    2. Students might use fraction strips or circles to show that $\frac{11}{12}$ leaves the smallest remainder in the whole unit, and therefore is the larger fraction.

Answer Key

28. 1. 25 students would get a portion with $\frac{1}{4}$ remaining.
    2. 17 students would get a portion.
    3. Students should use drawings of the sub divided into appropriate length portions.

29. 1.

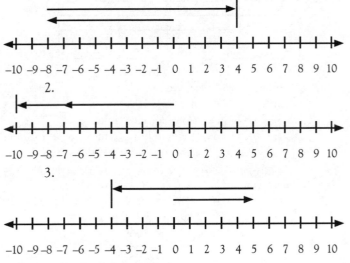

30. 1. $(-9) + 7 = (-2)$
    2. $(-4) + (-6) = (-10)$
    3. $7 + (-14) = -7$

31. 1. $5 + (-8) = (-3)$
    2. $(-6) + 10 = 4$
    3. $(-8) + 3 = (-5)$

32. 1. Chip board 1: 5 black chips; Chip board 2: 5 black chips and 9 white chips; Chip board 3: 4 white chips
    2. Chip board 1: 12 white chips; Chip board 2: 12 white chips and 7 black chips; Chip board 3: 5 white chips
    3. Chip board 1: 8 black chips; Chip board 2: 5 black chips crossed out or removed; Chip board 3 (or 2): 3 black chips

33. Chip board 1 should contain 6 black chips. Chip board 2 could show the 6 black chips with 6 "zeros" included. Chip board 3 would then show 6 white chips remaining after 12 black chips are removed. Thus $(-6) - (-12) = +6$.

34. 1. -18
    2. 33
    3. -21

4. −5
5. −1/4
6. 7
7. −2
8. −5
9. −7
10. −2.2
11. $-3\frac{4}{5}$
12. 187/10

35. Two weighings are needed. Choose any 6 coins. Put 3 on each side. If the 6 coins balance, weigh the remaining 2 coins. If the 6 coins don't balance, take the 3 on the heavy side and weigh any 2 of them. If they balance, you know the third coin is the heavy one. If they don't balance, then you have found the heavy coin.

36. 1,568 students; (791 × 2) − (2 × 7) = 1,568

37. $1 = \frac{4}{4}-(4-4)$; $2 = \frac{4}{4}+\frac{4}{4}$; $3 = \frac{(4+4+4)}{4}$;
    $4 = (4-4)\times 4+4$; $5 = (4\times 4+4)/4$; $6 = (4+4)/4+4$;
    $7 = (4+4)-4/4$; $8 = 4+4+4-4$; $9 = 4+4+4/4$;
    $10 = (44-4)/4$

38. Answers will vary. Possible solution table:

Step	8-gallon	5-gallon	3-gallon
1	8	0	0
2	3	5	0
3	3	2	3
4	6	2	0
5	6	0	2
6	1	5	2
7	1	4	3
8	4	4	0

39. Sybil: $39; Chuck: $21; Theodore: $12

40. 1. Indiana: 154.6/square mile; Wyoming: 4.7/square mile
    2. 5,375,570 people would have to leave.
    3. 3,926,254 people would have to move from Indiana to Wyoming to give the two states the same population density.

Part 2: Measurement

41. 1. 15 × 21 = 315/9 = 35 square yards × $7.32 = $256.20
    2. (15 + 21)2 = 72 feet; 72/8 = 9 pieces of molding × $134.00 = $1,206

Answer Key

42. The blob is between 19 and 20 square inches.
43. There are 12 different pentominoes. One has a perimeter of 10 units. All others have a perimeter of 12 units.
44. There are five unique rectangles: 1×48, 2×24, 3×16, 4×12, 6×8. The smallest perimeter is 28 units for the 6×8 rectangle. The largest perimeter is 98 units for the 1×48 rectangle. None are perfect squares.
45. The area is approximately 220 square feet. The area can be found by separating the figure into rectangles and triangles.
46. The original area on the grid is approximately 52 square units. The new area is approximately 34 square units. This represents a 34.6% decrease in area.
47. 1. One 12-inch pizza equals four 6-inch pizzas.
    2. One 18-inch pizza equals nine 6-inch pizzas.
48. An area pricing model would be best. Explanations will vary.
49. perimeter = 146.24 centimeters; area = 567.04 square centimeters
50. Each new tree has a basal area of $(3.5/2)^2 \times (3.14) = 9.621$ square inches. 592 square feet = 85,248 square inches. The Parks Department will need to plant 8,860 new trees.

51. For each triangle, $b = 5$, $h = 7$, and $A = 35$. They are not necessarily congruent; however, their areas are equal.
52. The 18-inch pizza provides the best value. 9-inch pizza: 63.6 square inches/10.5 = 6.06 square inches per $1.00; 12-inch pizza: 113.1 square inches/15 = 7.54 square inches per $1.00; 18-inch pizza: 254.5 square inches/19 = 13.39 square inches per $1.00
53. 1. $A = 100 - (100 \times 3.14/4) = 21.5$ square units
    2. $\{(100 \times 3.14) - 49 \times 3.14\}/4 = 40$ square units
54. 1. 4/10 of $260 = $104; 2. 9/40 of $260 = $58.50;
    3. 21/80 = $68.25; 4. 9/80 = $29.25
55. 1. 662 square inches
    2. 72 DVDs
    3. Yes; answers will vary.
56. 1. 3:1
    2. 3:1
    3. 9:1
    4. 27:1
57. $4.6 \times 0.75 = 3.45 \times 4 \times 365 = 5{,}037$ pounds/50 = 100.7 cubic feet
58. classroom = 600 cubic yards; 236 million tons $\times$ 2.5 cubic yards divided by 600 cubic yards = 980,000 classrooms

Daily Warm-Ups: Pre-Algebra, NCTM Standards

59. 1. The volume of the cone is ⅓ the volume of the cylinder.
    2. $1.00
60. Small container: $V = \pi r^2 h$; large container: $V = 4\pi r^2 h$. The girls get twice as much popcorn in the large container as in two small containers.
61. 1. volume of prism = 187.5 cubic inches; volume of cylinder = 147.26 cubic inches
    2. A comparable price for the prism container would be $4.19.
62. 1. Drawings will vary, but should reflect the pattern.
    2. 15 orange creams; $(R - 1)(C - 1)$ = number of orange creams for R rows and C columns of French mints.
63. Yes, there's room for 5 pairs. They require 15 to 20 square feet. 12 pairs would require 36 to 48 square feet. To provide the maximum space, Kanya and her father could increase the depth by 4 feet or the width by 6 feet.
64. 1. no
    2. yes
    3. no
    4. yes
    5. no
    Circumference of circle, c, is 2, but Rajah is given

a semicircle. Therefore, the circumference of the semicirle is πr. The rectangle has 3 sides to consider for the perimeter. The length is $2r$ and the height is r. Therefore, the perimeter of the rectangle $2h + 1 = 2r + 2r = 4r$. The total perimeter of the figure is $4r + \pi r$, or $r(4 + \pi)$.

65. b and c
    The volume of a cylinder is given by the formula $v = \pi r^2 h$ where πr^2 = base area. The height, h, of the cylinder can be represented by $x + y$. The volume is then expressed as $v = 45 (x + y) = 45x + 45y$.
66. area of rectangle = 75.36 square inches; total surface area = 89.52 square inches
67. 1. $1 \times 1 \times 12$; $1 \times 2 \times 6$; $1 \times 3 \times 4$; $2 \times 2 \times 3$
    2. 50 square feet; 40 square feet; 38 square feet; 32 square feet
    3. the carton that is $2 \times 2 \times 3$
    4. The carton that is $2 \times 3 \times 4$ needs 52 square feet of cardboard, 20 square feet more than a 12-ball carton.

68. surface area = 198 square inches; volume = 162 cubic inches

69. surface area = 261 square feet; volume = 390 cubic feet

    A possible net for the figure:

    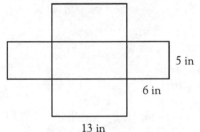

    5 in

    6 in

    13 in

70. 1. Answers will vary.
     2. volume of tall cylinder = 63.24 cubic inches
        volume of short cylinder = 76.96 cubic inches
        volume of tall prism = 49.67 cubic inches
        volume of short prism = 60.5 cubic inches
     3. The short cylinder is 16.46 cubic inches greater than the short prism.
     4. Answers will vary.

71. 1. 15 unit cubes
     2. 7 layers
     3. 105 in³; 142 in² (open box)

72. Volume of container = 509 cubic centimeters; volume of a tennis ball = 113 cubic centimeters; volume of air = 170 cubic centimeters. Notes will vary.

73. 1.

Time in hours	Northbound car's distance in miles	Eastbound car's distance in miles	Distance between cars in miles
1	60	50	78.1
2	120	100	156.2
3	180	150	234.3
4	240	200	312.4

   Each car's distance from the starting point is increasing by the amount of its rate of speed each hour. The distance between the cars is increasing by 78.1 miles each hour.

   2. 100² – 80² = 3,600; thus, the other car is traveling 60 mph.

   3.

   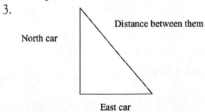

   North car

   Distance between them

   East car

74.

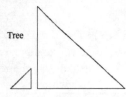

Tree

Shadow

$\frac{47.25}{3.5} = \frac{T}{3}$, $T = 40.5$; Eric's garage may be safe, but only by a foot and a half.

Part 3: Understanding Patterns, Relations, and Functions

75.

Side of pool	Number of border tiles
2	12
3	16
4	20
5	24
6	28
7	32

Tiles increase by 4: $B = 4S + 4$.
The graph is a straight line.

76. Maria may be counting all the tiles at the top and bottom of the border ($n + 2$) and then adding the remaining tiles on each side (n) for her expression. Carlos may be thinking that the border can be separated into 4 equal sets of ($n + 1$) sections.

77. Answers will vary. Sample answers: $4(S + 1)$; $2S + 2(S + 2)$; $(S + 2)^2 - S^2$; $4(S + 2) - 4$; $S + S + S + S + 4$

78. 1. 40 Popsicle sticks
2. explicit equation: $t_n = 3(n - 1) + 4$; recursive equation: $t_n = 4$, $t_n = t_{n-1} + 3$
3. Polly's pattern could not contain *exactly* 71 Popsicle sticks. The closest thing would be 70 Popsicle sticks—a chain of 23 parallelograms.

79. The fifth figure will have 15 dots. The sixth figure will have 21 dots. The equation is $T = \frac{n(n+1)}{2}$, where T is the value of the triangular number and n is the number of the term.

80. 1. This is a representation of square numbers.
2. 100 blocks
3. 10,000 blocks
4. $B = n^2$

Daily Warm-Ups: Pre-Algebra, NCTM Standards

Answer Key

81. You might want to encourage students to think about square numbers and triangular numbers.

$$P_n = S_n + T_n = n^2 + \frac{n(n-1)}{2} = \frac{n(3n-1)}{2}$$

82. 1. 40 pages
    2. 20 and 21; 41
    3. One possibility: N sheets of paper result in $4N$ pages. The sum of the page numbers on one side of 1 sheet is $4N + 1$.
    4. 820
    5. $2N$ represents the number of sides of each sheet. Thus, the sum of the pages = $2N(4N + 1)$.

83. One way to think about this situation is to imagine the stacked boxes are a staircase that is half of a rectangle, where n represents the base and $n + 1$ represents the height of the rectangle. The value $2B$ then would be equal to $n(n + 1)$. In other words, twice the number of boxes must equal the product of two consecutive integers. Solving for the equation $2B = n(n + 1)$ where $B = 45$ gives a quadratic where $n = -10$ or $n = 9$. Since Freya can't use a negative number of boxes, she should use 9 boxes on the bottom row.

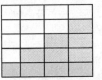

84. Paolo could make 29 pieces with 7 cuts.

Number of cuts	0	1	2	3	4	5	6	7	8	9	n
Number of pieces	1	2	4	7	11	16	22	29	37	46	$\frac{n(n+1)}{2}+1$

85. 1. 625; no middle number in row 50; n^2
    2. 6; 8; 78; $2(n-1)$
    3. 29; 41; $n^2 + n - 1$
    4. n^3
    5. $\left[\dfrac{n(n+1)}{2}\right]^2$

86. $D = 60t$. Letters will vary.

87.

U-Say	3	0	−4	1	2	5
I-Say	11	2	−10	5	8	17

Abby's rule: U = 3 • I + 2

88. 1. a. $y = 3^x$; As x increases, y increases by a factor of 3. There is a constant growth factor.
    b. missing values: (4, 81) (5, 243)
    c. The relationship is exponential.
2. a. $y = -x^2 + 4x$; As x increases, y increases by a decreasing amount. Second differences are constant.
    b. missing values: (4, 0) (5, -5)
    c. The relationship is quadratic.

89. 1. $x - y = 12$, $xy = 45$, thus $(12 + y)y = 45$, $y^2 + 12y - 45 = 0$, $(y + 15)(y - 3) = 45$, thus the values are 3 and 15, or -3 and -15

2.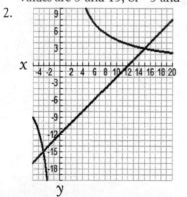

3. Students might choose values of x and y whose difference is 12 and test for the products. This strategy might not lead them to the negative values that satisfy this situation.

x	y	Product
13	1	13
14	2	28
15	3	45

4–5. Answers will vary.

90. 1. linear regression $(a + bx)$
    regEQ$(x) = 8.1875 + .458333x$
2–3. Answers will vary.

91. 1. 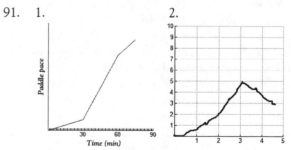   2.

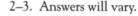

92. Graph A looks like it might be linear. Graph B looks like a curve that opens upward; as x increases, y decreases. Graph C does not seem to have a best-fit line or curve.

93. 1. quadratic
    2. linear
    3. quadratic
    4. exponential
    5. quadratic

94. 1. a. $y = 2x + 7$; As x increases, y increases by a constant amount. First differences are constant.
       b. missing values: (4, 15) (5, 17)
       c. The relationship is linear.

    2. a. $y = -x^2 + 11x$; As x increases, y increases by a decreasing amount. Second differences are constant.
       b. missing values: (4, 28) (5, 30)
       c. The relationship is quadratic.

Part 4: Understanding Algebraic Symbols

95.

Number of toothpicks	6	11	16
Number of hexagons	1	2	3
Perimeter of figure	6	10	14

$T = 5H + 1; P = 4H + 2$

96. 1. -3
    2. $-5\frac{1}{2}$
    3. 128
    4. 4.75
    5. 0
    6. 72
    7. 1
    8. 16
    9. 2
    10. 50

97. 1. 24
    2. 10
    3. 413
    4. 24
    5. 12.5
    6. -90

7. 3/4

8. 0

9. 0

10. −40

98. 1. $S = 4L + 4W + 4H = 4(L + W + H)$

2. $A = 2LW + 2LH + 2WH = 2(LW + LH + WH)$

99. 1. $2(5) + 2(4) + 4(2.5) = 28$ meters

2. $P = 2M + 2N + 4L = 2(M + N) + 4L$

3. Answers will vary.

100. Students might talk about undoing the steps and subtracting 17 from (or adding −17 to) both sides and then dividing by −4.

101. 1. 4

2. 6

3. 1

4. 0, 3.5

5. 0, 13

6. 0, −2

7. 0, 5

8. −2

9. −2

10. 0, 7

102. 1. $8x - 40$ or $4(2x - 10)$

2. $5x^2 - 6x + 13x - 10 = 5x^2 + 7x - 10$

3. $4x + 20 - 6 + 12x = 16x + 14$

4. $30 - 5x + 5x + 5 = 35$

5. $-5x^2 + 3x + 5$ or $5(1 - x^2) + 3x$

103. 1. $y = 8x - 21$; slope = 8, y-intercept = (0, 21)

2. $y = x - 1$; slope = 1, y-intercept = (0, −1)

3. $y = \frac{5}{3}x + 6$; slope = 5/3, y-intercept = (0, 6)

4. $y = 2.5x + 13$; slope = 5/2, y-intercept = (0, 13)

5. $y = 13.2x - 138.4$; slope = 66/5, y-intercept = (0, −138.4)

104. 1. $y = -3x + 11$

2. $y = -\frac{2}{3}x + \frac{5}{3}$

3. $y = -\frac{5}{4}x + \frac{17}{2}$

4. $y = -x + 1$

5. $y = x + 0$

105. 1. $y = \frac{3}{2}x + 2$

2. $y = 3$

3. The line passes through (1, 5) and (4, 6), thus $y = \frac{1}{3}x + \frac{14}{3}$.

4. $y = -x$

106. 1. $y = 2x + 3$
      2. $y = -3x + 5$
      3. $(y - 5) = \frac{1}{3}(x - 2)$
      4. $(y - 7) = \frac{2}{3}(x - 3)$

107. 1. $y = -300x + 350$ for x = cost and y = number sold
      2. The slope = -300, and the y-intercept = 350. For every $1.00 increase in price, the number sold decreases by 300; in addition, they can give away 350 brownies if they don't charge anything.
      3. 140 brownies
      4. about $0.17 for each brownie

108. Answers will vary.

109. 1. a line of best fit through the points
      2. a constant rate of change for the ordered pairs
      3. first-degree values for x and y
      4. Answers will vary.

110. Answers will vary. Sample answer: If you know the slope and y-intercept, you can substitute the values into the equation $y = mx + b$ or $y = a + bx$. The slope can be determined from the graph by finding the change in y divided by the change in x. The y-intercept will be the point where $x = 0$ in the table and where the line crosses the y-axis on the graph.

111. Answers will vary. Sample answer: The rate of change is the value found when the difference in two y values is divided by the difference in the corresponding two x values, either from the table or from the graph using any two points. In the equation, the rate of change is the value m or b when the equation is written in the form $y = mx + b$ or $y = a + bx$. The y-intercept is the value $(0, x)$.

112. 1. $C = f(T) = T - 37$

2. **Cricket Chirps**

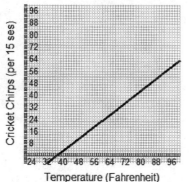

3. Answers will vary.

113. Slope is $\frac{212-32}{100-0} = \frac{180}{100} = \frac{9}{5}$, thus $y = \frac{9}{5}x + 32$.
Note: y = Fahrenheit and x = Celsius.

114. 1. Eric weighs about 104 pounds.

2. If Amber and Aleah continue to sit at 4 feet, then $130 \times 4 = 104 \times D$, and $D = 5$ feet. If the seesaw is long enough, then Eric can balance them. Otherwise, he cannot.

115.

Number of subscriptions sold	Weekly pay (dollars)
1	145
2	170
3	195
6	270
10	370
12	420
15	495

2. $W = 25S = 120$; $S = (W - 120)/25$
3. 20 subscriptions

116. Answers will vary. Sample answers:
1. C might represent the original cost of a coat with a discounted price of $37.49. C = $49.99.
2. The sale price of a shirt might be $28.00 after a 20% discount. C represents the original cost of the shirt. C = $35.00

117. $70m + 100 > 500$; $70m > 400$; $m > 5.7$. School started 6 months ago.

118. 1. $\dfrac{1}{x^n}$

2. y^{r-s}

3. x^{ab}

4. $a^n b^n$

5. $\dfrac{d^n}{t^n}$

6. x^{-n}

7. n^{p+q}

8. $\sqrt[n]{y}$

119. 1. $x + y = 10$; $x + 2y = 8$; $(12, -2)$
2. $x + 2y$; $x + y = 15$; $(10, 5)$
3. $x - y = 2$; $2x - 2y = 4$; Any values for x and y that differ by 2 will work in this puzzle. There are infinite solutions.

120. Word puzzles will vary.

1. $(2, 1)$
2. $(-2, 4)$
3. not possible

121. 1. correct
2. $5x$
3. $6n^2$
4. $6xy$
5. $a + b$
6. $n + 5/n$ or $1 + 5/n$
7. correct
8. $2x^3$
9. correct
10. $5a + 7b$

122. Students' choices and rationales may vary.
1. $(-4, -5)$
2. $(-6, 4)$
3. $(-20, 10)$
4. $(3.6, 2.4)$

Daily Warm-Ups: Pre-Algebra, NCTM Standards

123. Answers will vary.
   1. Sample answer: to find an ordered pair that satisfies both equations

   2. Sample answer: Graph both lines on the same set of axes, and locate the intersection point of the two lines.

   3. Sample answer: Substitute values of x into both equations. Look for identical y values for the same value of x.

   4. Sample answer: solve algebraically by substitution or elimination

   5. Sample answer: If the values in the table have the same constant rate of change, if the lines are parallel, or if the slopes are the same but have different y-intercepts, then there is no solution.

Part 5: Developing Algebraic Thinking Through and for Mathematical Modeling

124. There is a difference of 2 between the two products. The numbers can be represented as n, $n + 1$, $n + 2$, $n + 3$; $n(n + 3) = n^2 + 3n$ represents the outer products. $(n+1)(n+2) = n^2 + 3n + 2$ represents the inner products, thus the inner product will always be 2 more than the outer product.

125.

Number of purple rods	1	2	3	4	5	6	n
Surface area	18	28	38	48	58	68	$10n + 4$

126.

Number of purple rods	1	2	3	4	5	6	n
Surface area	18	30	42	54	66	78	$18 + 12n + 2$

127. Students may need to be reminded to convert 12.5 feet to 381 centimeters. Students' equations may vary depending on the strategy used. This data and graph leads to an answer of 23 rubber bands.

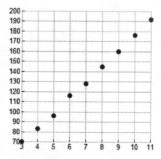

Answer Key

128. $H = 120$

Number of students (n)	Number of handshakes (h)
1	0
2	1
3	3
4	6
5	10

$H = \dfrac{n(n-1)}{2}$ where n = number of people

Handshakes

129. 33 people provide the greatest income for EVA.

Number of people	10	14	18	22	26	30	31
Cost per person	200	184	168	152	136	120	120
Total income	2000	2576	3024	3344	3536	3600	3720

Number of people	32	33	34	35	36	37	38
Cost per person	120	120	104	104	104	104	88
Total income	3840	3960	3536	3640	3744	3848	3344

130. 1–2.

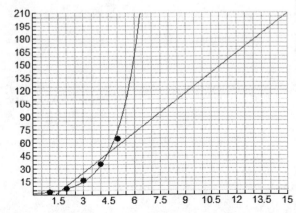

Daily Warm-Ups: Pre-Algebra, NCTM Standards

3. Linear regression $(a + bx)$
   regEQ$(x) = -20.3 + 15.3x$

4. Using a linear model as a predictor indicates that Linn might have 209 mice after 15 months.

5. Using an exponential graph as a predictor indicates that there may be 200 mice by 7 months.

131. linear regression $(a + bx)$
   regEQ$(x) = 59.0067 + -.003567x$

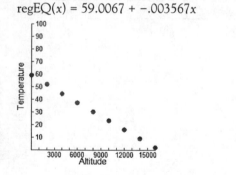

132. 1. $A = (n(120 - 2n)$ expresses the area of the pen when one side is against the barn, where n = width.
   $(n = 30; m = 60)$

2. $n = 40; m = 40; A = 1,600$

3. The maximum area using the barn is 1,800 square feet.

4. Minimum values will vary depending on students' decisions on how short a width can be. One possible answer is 118 square feet using the barn.

5. The maximum area without using the barn is 900 square feet. Minimum areas will vary; one possible answer is 59 square feet.

133.

Side of square cutout(s)	Length of box (12 − 2s)	Width of box (9 − 2s)	Volume of box s(12 − 2s)(9 − 2s)	Surface area of box
1	10	7	70 in³	104 in²
2	8	5	80 in³	92 in²
3	6	3	54 in³	72 in²
4	4	1	16 in³	44 in²

Greatest volume results from 2-inch cutouts. Smallest volume is from 4-inch cutouts.

134. A table and graph might look like this:

Day	Salary	Uncle's commission
1	1,000	100
2	1,800	200
3	3,200	400
4	5,600	800
5	9,600	1,600
6	16,000	3,200
7	25,600	6,400
8	38,400	12,800
9	51,200	25,600
10	51,200	51,200
11	0	102,400
12	−102,400	204,800
13	−204,800	409,600

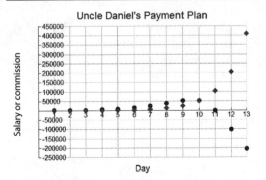

Uncle Daniel's Payment Plan

It might not be wise to continue to work for Uncle Daniel after day 10.

135. The unit cube has 8 corners, 6 faces, and 12 edges.
$T = n^3$

L1	L2
1	1
2	8
3	27
4	64
5	125
6	256
7	343
8	512
9	729
10	1000

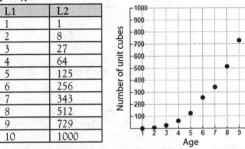

136. 1.

Side length	3 faces	2 faces	1 face	Zero faces
2	8	0	0	0
3	8	12	6	1
4	8	24	24	8
5	8	36	54	27
6	8	48	96	64
7	8	60	150	125
8	8	72	216	216
9	8	84	294	343
10	8	96	384	512
15	8	156	1014	2197

2. 3 faces located in corners; 2 faces located on edges; 1 face located on faces other than those listed above; 0 faces located on the interior

3. For all cubes, 3 faces is 8; 2 faces: $N = 12(s - 2)$; 1 face: $N = 6(s - 2)^2$; zero faces: $N = (s - 2)^3$.

137.
1. $2,030.00
2. $2,060.45
3. $2,090.90
4. After 15 months, his unpaid balance will be greater than $2,500.

138.
1. 23 toothpicks
2. explicit equation: $t_n = 2n + 1$; recursive equation: for $t_1 = 3$, $t_n = t_{n-1} + 2$
3. There would be 35 connected triangles.

139.
1. 15 moves
2.

Number of each color	1	2	3	4	5	6	7	8	9	10
Number of moves	3	8	15	24	35	48	63	80	99	120

3. $M = n(n + 2)$

140. Solutions will vary depending on students' choices. One graph and linear regression equation are provided for the data for both sexes.

Linear regression $(ax + b)$
regEQ$(x) = .229738x + -381.817$
$a = .229738$
$b = -381.817$
$r = .981234$
$r^2 = .96282$
For this graph and equation, a student might predict that a person born in 2020 would have a life expectancy of 82.26 or 82.3 years.

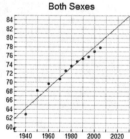

Both Sexes

141.
1. b
2. c

142.
1. c
2. a

143. Students should choose to justify choice *c* as the proper graph.

Answer Key

144. 1. between 4 and 6 seconds
      2. about 13 mps; 48 mps
      3. from 1 to 4 seconds into the ride; 6 to 7 seconds
         into the ride
      4. The velocity increased rapidly. They stayed at 55
         mps for 1 second, then decreased between 7 and 8
         seconds, increased to about 53 mps, and then began
         to decrease again.

145. 1. 2.03 million per year from 1900 to 2000 is less than
         the 2.8 million per year from 1990 to 2000.
      2. Highest rate of change: 1950 to 1960 or 1990
         to 2000. Lowest rate of change: 1930 to 1940.
         Calculate slope to find rates of change. Factors
         might include the postwar baby boom, the Great
         Depression, and so forth.
      3. Answers will vary. Sample answer: 325,000,000

146. $C_1 = 20.00 + 10m$; $C_2 = .40m$

Minutes	Horizon	Best Talk
0	0	20
10	4	21
20	8	22
30	12	23
40	16	24
50	20	25
60	24	26
70	28	27
80	32	28
90	36	29

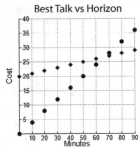

Best Talk vs Horizon

In the graph, Best Talk is represented by the diamond
shapes and Horizon is represented by the circles.
Students may argue that Horizon is the best plan until
Aisha exceeds 65 minutes in a month.

Daily Warm-Ups: Pre-Algebra, NCTM Standards

147.

Original square		Rectangular plot			Difference in areas
Side length	Area	Length	Width	Area	
4	16	7	1	7	9
5	25	8	2	16	9
6	36	9	3	27	9
7	49	10	4	40	9
8	64	11	5	55	9

This is not a fair trade with respect to area. The rectangle will always be 9 square meters smaller than the square. The area of a square = n^2. The area of a rectangle = $(n+3)(n-3) = n^2 - 9$.

148.   1. $T = 2,000 + 399m$; $T = 1,600 + 420m$
2. At the end of 1 year, Joe's car will have cost $6,788 and Hank's car will have cost $6,640. At the end of 2 years, Joe's car will have cost $11,576, and Hank's car will have cost $11,680.
3. Answers will vary. Students could make an argument for either plan.

149.   1.

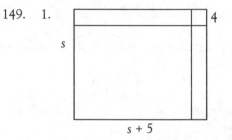

2. area of square = s^2, area of rectangle = $(s+5)(s-4)$

s	Area of square	Area of rectangle
5	25	10
10	100	90
15	225	220
20	400	400
25	625	630
30	900	910
35	1,225	1,240

3. For $s > 20$, the area of the rectangle will be greater than the area of the square.
4. For $s < 20$, the area of the rectangle will be less than the area of the square.
5. For $s = 20$, the areas will be equal.
6. Answers will vary.

Part 6: Analyzing Change in Various Contexts

150.   1. slope = −600/2 or −300
        2. rate of change = 300 mps downward
        3. The domain for the line segment is $1 \leq s \leq 3$. The range for the line segment is $1,800 \leq A \leq 2,400$.

151.   1. $C = 4.75(n) + 1.00$, C = total cost, n = number of packages
        2.

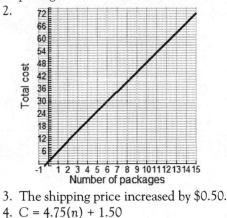

        3. The shipping price increased by $0.50.
        4. $C = 4.75(n) + 1.50$

152.   Graphs will vary depending on students' interpretations of the situations. Most students will create line or bar graphs. Discuss the appropriateness of a continuous or discrete graph for these models.

153.   Answers will vary. A graph of the combined data is given. An argument could be made for either plan. If Rosa's family rents 25 DVDs per month, but they only watch 2 movies per month on average, the Deluxe company is a better choice for her family.

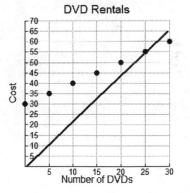

154. 1.

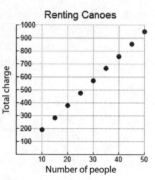

It would not make sense to connect the points on the graph. The data points are discrete, not continuous.

2. As the number of campsites increases, the fee increases a constant amount. The graph shows this with a slope up and to the right.

155.

Number	Total Charges
10	190
15	285
20	380
25	475
30	570
35	665
40	760
45	855
50	950

Renting Canoes

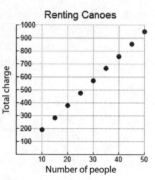

156. 1. Variables are time and number of step-ups.

2. As time increases, the number of step-ups in each 10-second interval decreases.

y	# step ups	11	19	29	39	48	52	59	62	65	69	72	75
x	time in seconds	10	20	30	40	50	60	70	80	90	100	110	120

3. Estimates: (25, 25), (65, 55); explanations will vary.

Answer Key

157. The weight increases more each month in the first year than in the second year. The graph looks like it might fit a parabolic curve for $0 \leq x \leq 25$.

Quadratic regression
regEQ$(x) = -.035955x^2 + 1.63514x + 7.82424$

158. At 80 miles per hour, the stopping distance would be close to 500 feet. Explanations will vary.

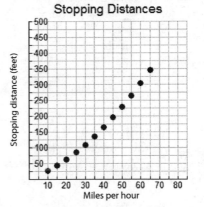

159. 1. A table of values would have increasing values of x and simultaneously increasing values of y. The graph would slope up and to the right.
2. This situation would have x values increasing and y values decreasing, so the slope of the graphed line would be down and to the right.

Daily Warm-Ups: Pre-Algebra, NCTM Standards

3. The points lie on a straight line when there is a common slope between each set of points.
4. The points could be connected if the data expressed represents a continuous change in values.

160. 1. growth; R = 3; 300% increase
2. growth; R = 1.1; 110% increase
3. decay; R = .6; 60% decrease
4. decay; R = .2; 20% decrease
5. decay; R = .38; 38% decrease

161. 1. y-intercept: (0, 16); x-intercepts: (−4, 0), (4.3, 0). Other answers will vary.
2. y is increasing over the interval (−5, 0).
3. No, the rate of change is not constant. The change over (0, 2) is −1 , over (2, 4) is −7/2, and over (4, 5) is −7.

162.

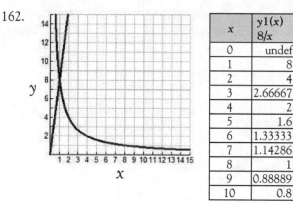

x	$y1(x)$ $8/x$	$y2(x)$ $8x$
0	undef	0
1	8	8
2	4	16
3	2.66667	24
4	2	32
5	1.6	40
6	1.33333	48
7	1.14286	56
8	1	64
9	0.88889	72
10	0.8	80

least values: (10, 0.8) and (0, 0); greatest values: (1, 8) and (10, 80); intersection: (1, 8)

163. 1. speed = 1.25 kilometers per minute
2. $d = 1.25t$
3.

Time	20	40	60	80	100
Distance	24.5	50	75	100	125

Answer Key

4. 24.5 kilometers
5.

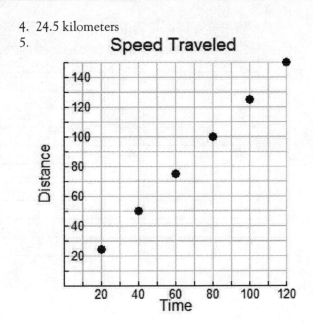

Speed Traveled

164. The graphs of the equations are quadratic, indicated by the constant values in the second-difference columns. The tables are shown below.

1.

x	y = 3x(x+2)	Difference in y's	Second Difference
–5	45	45 – 24 = 21	21 – 15 = 6
–4	24	24 – 9 = 15	15 – 9 = 6
–3	9	9 – 0 = 9	6
–2	0	3	6
–1	–3	–3	6
0	0	–9	6
1	9	–15	6
2	24	–21	6
3	45	–27	6
4	72	–33	
5	105		

2.

x	$y = 2x - x^2$	Difference in y's	Second Difference
−5	−35	−11	2
−4	−24	−9	2
−3	−15	−7	2
−2	−8	−5	2
−1	−3	−3	2
0	0	−1	2
1	1	1	2
2	0	3	2
3	−3	5	2
4	−8	7	
5	−15		

3.

x	$y = x^2 + 5x + 6$	Difference in y's	Second Difference
−5	6	4	2
−4	2	2	2
−3	0	0	2
−2	0	−2	2
−1	2	−4	2
0	6	−6	2
1	12	−8	2
2	20	−10	2
3	30	−12	2
4	42	−14	
5	56		

Part 7: Data Analysis and Probability

165. 1.

Dice 1/Dice 2	1	2	3	4	5	6
1	2	3	4	5	6	7
2	3	4	5	6	7	8
3	4	5	6	7	8	9
4	5	6	7	8	9	10
5	6	7	8	9	10	11
6	7	8	9	10	11	12

2. P(2) = 1/36; P(3) = 2/36; P(4) = 3/36; P(5) = 4/36; P(6) = 5/36; P(7) = 6/36; P(8) = 5/36; P(9) = 4/36; P(10) = 3/36; P(11) = 2/36; P(12) = 1/36

3. P(7 or 11) = 8/36; P(not 7 or 11) = 28/26

166. 1. P(tan) = 2/10
2. P(black or blue) = 6/10
3. P(tan second try) = 1/9
4. P(any pair) = 18/90

167. 1. N(H) D(H)
2. 25%
3. 50%
4. 75%

168. There are 72 different cones possible without repeated flavors, or 81 different cones possible if flavors are repeated. There are 36 different cones possible if order doesn't matter for different scoops, and 45 different cones if doubles are allowed.

169. 1. 5/8
2. 2/8
3. 0
4. 4/8
5. 4/8
6. 1/8

170. 1. Player A wins 4/8 of the time, and Player B wins 3/8 of the time (or 4/7 and 3/7 if the 4 spin is ignored).
2. The game is not fair. Answers will vary. One simple change could be that Player A wins on an even number and Player B wins on an odd number.

171. Answers will vary. For example, Player A scores 1 when the spinner lands on 1, 2, 3, or 4, and Player B scores 1 when the spinner lands on 5 or 6. Neither would score on a 7 or 8.

172. The circle is divided into 60% and 40% sections. Spinning could simulate an attempted shot. If the spinner lands in the 60% portion on the first shot, the shooter would shoot again, thus the spinner would be spun again. Thus an event would be 0 points, 1 point, or 2 points scored on each appearance at the foul line.

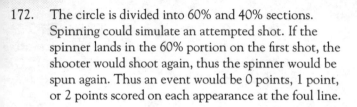

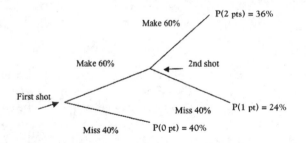

Make 60%   P(2 pts) = 36%

Make 60%   2nd shot

First shot

Miss 40%   P(1 pt) = 24%

Miss 40%   P(0 pt) = 40%

173. 1. $30,266/person
2. 191,016,307 median incomes
3. 232,332 Fitzpatrick incomes

174. 1. R = 1,539
2. 2004 mean = 527 million; 1980 mean = 357 million; the difference = 170 million
3. The United States produced 22% of emissions in 2004.

175. 1. 41%
2. $1.40/loaf
3. The 2005 price is 30% greater. If the trend were continuing, one might assume that the percent increase would be closer to 20%.

176. The spinner should have regions with the following approximate angle measures: red = 90°, blue = 135°, yellow = 60°, green = 45°, and purple = 30°. The arrangement of the circle segments doesn't matter.

177. 1. $P(T) = 1/2$
     2. $P(quarter) = 0$; $P(nickels) = 0$
     3. Answers will vary.
178. This is not a very good fund-raiser. The class could potentially pay out the same amount they take in because the probability of winning is 1/4. For 20 rolls, the player could expect to win 5 times for $20. The game could be considered fair from a money-gained-or-lost point of view.
179. It is not likely that this is a 50% probability. Students might draw diagrams that show by geometric probability that the areas of winning and losing are not half and half.
180. 1. $P(H) = 1/4$
     2. 3 hearts
     3. 3 hearts
     4. theoretical probability